Insulating Concrete Forms Construction Manual

Insulating Concrete Forms Construction Manual

Pieter A. VanderWerf
W. Keith Munsell

McGraw-Hill

New York San Francisco Washington, D.C. Auckland Bogotá
Caracas Lisbon London Madrid Mexico City Milan
Montreal New Delhi San Juan Singapore
Sydney Tokyo Toronto

McGraw-Hill

A Division of The McGraw·Hill Companies

© 1996 by **Portland Cement Association**.
Published by The McGraw-Hill Companies, Inc.

Printed in the United States of America. All rights reserved. The publisher takes
no responsibility for the use of any materials or methods described in this book,
nor for the products thereof.

pbk   1 2 3 4 5 6 7 8 9 FGR/FGR 9 0 0 9 8 7 6 5

Library of Congress Cataloging-in-Publication Data
VanderWerf, Pieter A.
    Insulating concrete forms construction manual / by Pieter A.
VanderWerf & W. Keith Munsell.
      p.   cm.
    Includes index.
    ISBN 0-07-067032-3 (pbk.)
    1. Concrete houses—Design and construction. 2. Insulating
concrete forms. I. Munsell, W. Keith.
    TH4818.C6V357 1995
    693'.5—dc20   95-39611
                                                      CIP

McGraw-Hill books are available at special quantity discounts to use as premiums
and sales promotions, or for use in corporate training programs. For more
information, please write to the Director of Special Sales, McGraw-Hill, 11 West
19th Street, New York, NY 10011. Or contact your local bookstore.

Acquisitions editor: April D. Nolan
Editorial team: Robert E. Ostrander, Executive Editor
              Sally Anne Glover, Book Editor
Production team: Katherine G. Brown, Director
              Lisa M. Mellott, Coding
              Jeffrey Miles Hall, Computer Artist
              Wanda S. Ditch, Desktop Operator
              Linda L. King, Proofreading
              Jodi L. Tyler, Indexer
Design team: Jaclyn J. Boone, Designer
              Katherine Lukaszewicz, Associate Designer   0670323
                                                                       GEN3

Contents

Introduction

This manual is a step-by-step contractor's guide to using insulating concrete forms (ICFs). ICFs are hollow blocks, panels, or planks made of rigid foam that are erected and filled with concrete to form the structure and insulation of exterior walls. Figures I-1A through I-1D show the stages of construction. ICFs are growing rapidly in popularity because they are cost-competitive with frame construction, easy to learn and use, and environmentally friendly, yet they deliver a higher-quality product: more energy efficient, comfortable, durable, stronger, quieter, and more resistant to natural elements and disasters. They have been used mostly for houses, up to a full basement plus three stories. However, they are also suitable for commercial construction and can go much higher with proper design. ICFs are currently used to build walls only, although several manufacturers are designing additional forming components that will allow the construction of attached concrete floors at the same time.

I-1A *Setting insulating concrete forms.*

I-1B *Completed formwork.*

I-1C *Pumping in concrete.*

I-1D *The completed home.*

There were about 20 different brands of ICFs in the United States as of the summer of 1995, and 4–5 new ones appearing each year. Information helpful to choosing the right brand for you is included in *The Portland Cement Association's Guide to Concrete Homebuilding Systems* (McGraw-Hill, 1995). Some are well documented, but they are so new and changing so fast that contractors are finding they need additional information and support.

This manual is the first to give step-by-step instructions for constructing houses out of the full range of available ICFs. It includes systems sold and used to build houses in the United States as of the summer of 1995. To gather the information, we started by reading the manufacturers' manuals. Then we found the most experienced, efficient ICF builders in the country, hounded them for details of their operations, and closely watched their crews as they worked. This manual gives the details of how the best builders and crews do the job, from initial planning through all phases of building. When different people did things differently, we picked the methods that were most preferred so we could keep the instructions simple. While we cannot make personal guarantees about the correctness or effectiveness of the methods presented here, they are all used by successful ICF builders.

This manual is not a substitute for the manufacturer's instructions. Always read and follow the recommendations of the manufacturer of the ICF system you are using. The directions here are for initial planning purposes and to serve as a guide when no manufacturer instructions are available.

This manual is based on the reported practices of actual builders. It is intended for the use of building professionals who are competent to evaluate the significance and limitations of these reported practices and who will accept responsibility for use of the material it contains. The Portland Cement Association is not responsible for application of the material contained in the manual or for the accuracy of material provided by sources other than PCA.

Note that there are also some worthy ICF systems sold only in Canada. Although we do not cover these specifically in this manual, most of the content should be applicable to them also. Appendix A contains the manufacturers' names and addresses for readers interested in using one of these systems.

The different ICFs cluster into a few logical types. Chapter 1, "Types of systems," describes them all. You can skip this if you already know which system you will be using and want to get right to the instructions. Chapter 2, "Tools and materials," is important because some products used in ICF construction are unusual in low-rise building, and some other products that are common get used in unusual ways. This chapter explains what the important tools and materials are, where to get them, and how to use them. It is a good idea to read this in advance. In case you skip it, it is organized so you can find the explanations you need if the manual confuses you with unfamiliar instructions like "use a hot wire" or "add plasticizer to your concrete."

Chapters 3–9 each cover one phase of construction. Some things (for example, how to attach wallboard) must be done differently with different ICF systems. And some things must be done differently, depending on your particular conditions (weather, type of foundation, and so on). In these cases we describe alternative methods to cover every brand and common situation. But otherwise we usually describe only the most preferred way of doing each task. Readers who are interested in considering the many alternative construction techniques not covered here can find them in the *Design & Construction Handbook for Stay-in-Place Forming Systems* (also published by McGraw-Hill, due in 1996).

The appendix is a directory of product and information sources. It contains the addresses and phone numbers of the companies that sell the ICF systems, the companies that sell the tools and materials you might need, and the people who can answer the questions you might have. It is organized so that you can quickly find the contacts you want. Good luck and good building.

1

Types of systems

Table 1-1 lists the ICFs we know to be sold and used to build houses in the United States as of the summer of 1995. They all work on the same principle (filling foam forms with concrete), but they have three differences that are important for the contractor: the size of the form units and the ways they connect to one another, the shape of the cavities into which the concrete goes, and whether the formwork has surfaces that can be fastened to with a screw or nail.

When we divide the systems according to these three areas of difference, we get eight categories of systems, as explained below. In addition, some systems have some unique feature that can be important, so we describe these things, too.

Three key differences

The first key difference is that the systems vary in their unit sizes and connection methods. We divide them into *panel*, *plank*, and *block* systems. Figure 1-1 contains a diagram of each.

Panel systems are the largest units, as big as 4 feet by 8 feet. This allows a lot of wall area to be erected in one step, but may require more cutting. The panels have flat edges and are connected to one another with extra fasteners: glue, wire, or plastic channel.

Plank systems consist of long (usually 8 feet), narrow (8 or 12 inches) planks of foam held a constant distance apart by steel or plastic ties. The planks have notched, cut, or drilled edges that the ties fit into. In addition to spacing the planks, the ties connect each course of planks to the one above and below.

Block systems include units ranging from standard concrete block size (8-x-16 inches) to a much larger 16 inches high by 4 feet long. Along their edges are teeth or tongues and grooves for interlocking; they stack without separate fasteners on the same principle as children's Lego blocks.

The second key difference is the shape of the cavities. Each system has one of three distinct cavity shapes: *flat, grid,* or *post-and-beam.* These produce different shapes of concrete beneath the foam, as shown in Fig. 1-2. Flat cavities produce a concrete wall of constant thickness, just like a conventional poured wall made with plywood or metal forms. Grid cavities are "wavy," both hori-

Table 1-1. Available ICF Systems[1]

	Dimensions[2] (width × height × length)	Fastening surface	Notes
Panel systems			
Flat panel systems			
R-FORMS	8" × 4' × 8'	Ends of plastic ties	Assembled in the field; different lengths of ties available to form different panel widths.
Styroform	10" × 2' × 8'	Ends of plastic ties	Shipped flat and folded out in the field; can be purchased in larger/ smaller heights and lengths.
Grid panel systems			
ENER-GRID	10" × 1'3" × 10'	None	Other dimensions also available; units made of foam/cement mixture.
RASTRA	10" × 1'3" × 10'	None	Other dimensions also available; units made of foam/cement mixture.
Post-and-beam panel systems			
Amhome	9⅜" × 4' × 8'	Wooden strips	Assembled by the contractor from foam sheet. Includes provisions to mount wooden furring strips into the foam as a fastening surface.
Plank systems			
Flat plank systems			
Diamond Snap-Form	1' × 1' × 8'	Ends of plastic ties	
Lite-Form	1' × 8" × 8'	Ends of plastic ties	
Polycrete	11" × 1' × 8'	Plastic strips	
QUAD-LOCK	8" × 1' × 4'	Ends of plastic ties	
Block systems			
Flat block systems			
AAB	11.5" × 16¾" × 4'	Ends of plastic ties	
Fold-Form	1' × 1' × 4'	Ends of plastic ties	Shipped flat and folded out in the field.

	Dimensions[2] (width × height × length)	**Fastening surface**	**Notes**
GREENBLOCK	10" × 10" × 3'4"	Ends of plastic ties	
SmartBlock Variable Width Form	10" × 10" × 3'4"	Ends of plastic ties	Ties inserted by the contractor; different length ties available to form different block widths.
Grid block systems with fastening surfaces			
I.C.E. Block	9¼" × 1'4" × 4'	Ends of steel ties	
Polysteel	9¼" × 1'4" × 4'	Ends of steel ties	
REWARD	9¼" × 1'4" × 4'	Ends of plastic ties	
Therm-O-Wall	9¼" × 1'4" × 4'	Ends of plastic ties	
Grid block systems without fastening surfaces			
Reddi-Form	9⅝" × 1' × 4'	Optional	Plastic fastening surface strips available
SmartBlock Standard Form	10" × 10" × 3'4"	None	
Post-and-beam block systems			
ENERGYLOCK	8" × 8" × 2'8"	None	
Featherlite	8" × 8" × 1'4"	None	
KEEVA	8" × 1' × 4'	None	

[1] All systems are listed by brand name. For the names and addresses of their manufacturers, see appendix A.

[2] "Width" is the distance between the inside and outside surfaces of foam of the unit. The thickness of the concrete inside will be less, and the thickness of the completed wall with finishes added will be greater.

zontally and vertically. If the forms are stripped away, the concrete looks like a breakfast waffle. Post-and-beam cavities are so named because you fill cavities with concrete only every few feet horizontally and vertically. In the most extreme post-and-beam systems, there is a 6-inch diameter concrete "post" every 4 feet and a 6-inch concrete "beam" at the top of each story.

Note that no matter what the cavity shape is, all systems also have "ties." These are the crosspieces that connect the front and back layers of foam. When the ties are metal or plastic, they do not affect the shape of the concrete much. But in some of the grid systems, they are foam and are much larger, forming breaks in the concrete about 2 inches in diameter every foot or so. Figure 1-3 shows the differences.

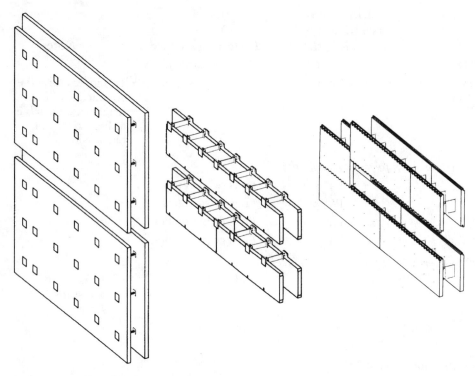

1-1 *Diagrams of ICF formwork made with the three basic units: panel (left), plank (center), and block (right).*

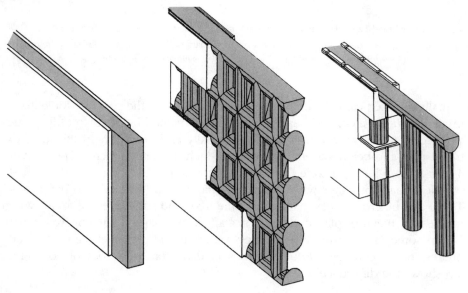

1-2 *Cutaway diagrams of ICF walls with the three basic cavity shapes: flat, grid, and post-and-beam.*

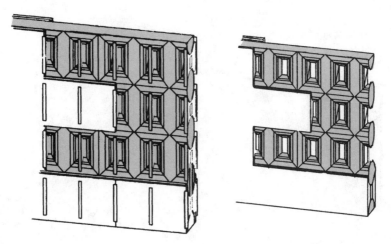

1-3 *Cutaway diagrams of ICF grid walls with steel/plastic ties and foam ties.*

The third key difference is that many of the systems also have a *fastening surface*—some other material (not foam) embedded into the units that crews can sink a screw or maybe a nail into, similar to fastening to a stud. Often this surface is just the ends of the ties. But other systems have no embedded fastening surface. Their units are all foam, including the ties. This generally makes them simpler and less expensive, but requires crews to take extra steps to connect interior wallboard, trim, exterior siding, and so on to the walls.

Eight categories

The different brands of ICF systems mix and match the key features (size and type of connection, shape of cavities, having or not having a fastening surface) in different ways. Table 1-1 lists the eight different combinations that are currently sold, the brands that fall under each one, and some key facts about each brand. Figures 1-4 through 1-11 contain diagrams of all eight of these different categories of systems.

Unique features

As noted in Table 1-1, a few brands have important additional features that are unique. R-FORMS is a flat panel system that you assemble out of ties and stock foam sheet yourself, as in Fig. 1-12. Tie lengths are adjustable; you form a longer or shorter tie depending on whether you want a thick or thin wall.

The units of the two grid panel systems (ENER-GRID and RASTRA, which are similar) are not made of pure foam, but of a mixture of cement and foam beads. Figure 1-13 contains a photo. This makes them different in several ways.

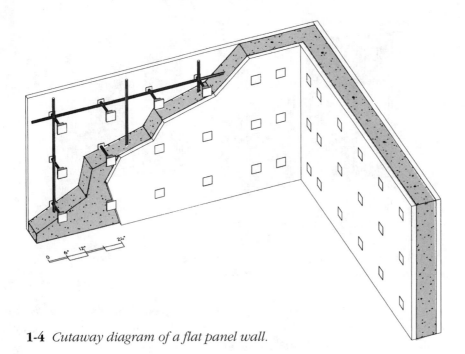

1-4 *Cutaway diagram of a flat panel wall.*

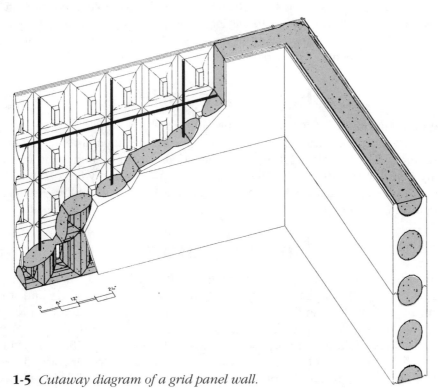

1-5 *Cutaway diagram of a grid panel wall.*

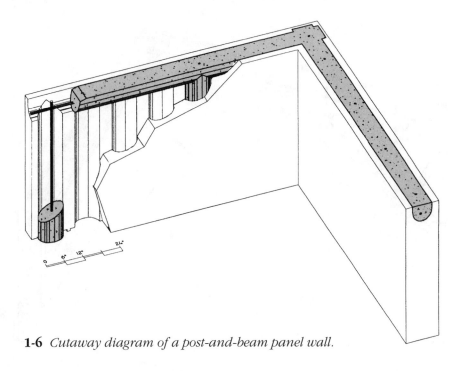

1-6 *Cutaway diagram of a post-and-beam panel wall.*

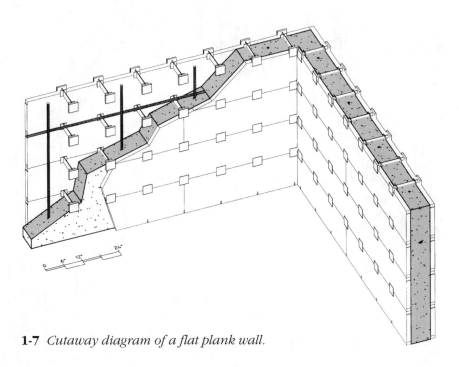

1-7 *Cutaway diagram of a flat plank wall.*

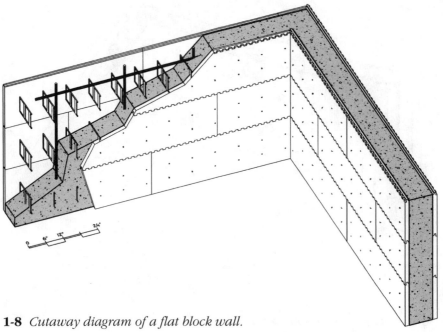

1-8 *Cutaway diagram of a flat block wall.*

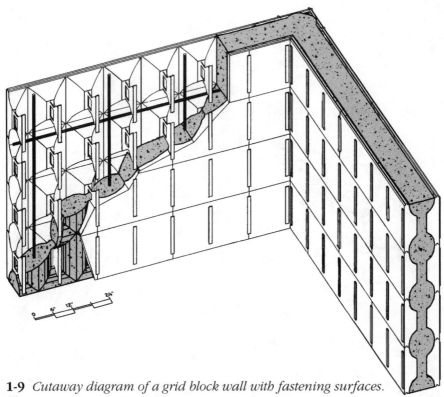

1-9 *Cutaway diagram of a grid block wall with fastening surfaces.*

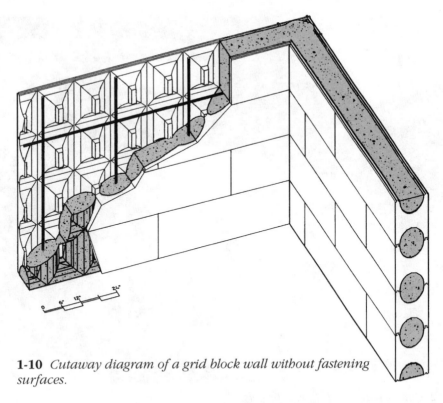

1-10 *Cutaway diagram of a grid block wall without fastening surfaces.*

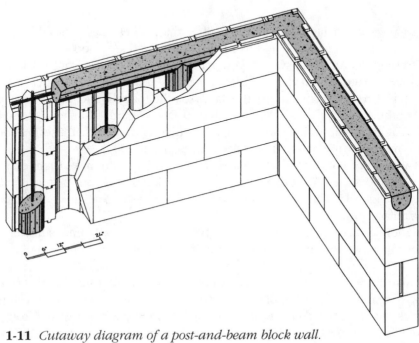

1-11 *Cutaway diagram of a post-and-beam block wall.*

R-FORMS, Inc.

1-12 *Site assembly of R-Forms panels.*

The in units are a little heavier and stiffer than other ICFs. They are a little harder, so they cut and drill less like foam and more like wood.

In using the Amhome system, you make your own panels from stock 8-inch foam sheet. You cut out the cylinders for posts and beams with special tools. You can also cut grooves into the faces of the panels and insert and glue wooden furring strips to act as a fastening surface. Figure 1-14 shows these features. The Amhome wall panels are part of a larger system that also includes a high-R roof structure made of wooden I-joists and foam.

The units of Fold-Form, a flat block system, collapse to make them easier to transport and store. You open them up on the job site before use, as demonstrated in Fig. 1-15. The units of Styroform, a panel system, collapse and fold out in a similar manner.

Reddi-Form, normally an all-foam block with no fastening surface, also comes in a new version with slots for the insertion of plastic strips. These serve as a fastening surface.

SmartBlock comes in two varieties, both shown in Fig. 1-16. The first is a flat block system that you assemble in the field out of ties and flat foam rectangles. The ties slide into two rectangles, connecting them into a block with a front and back face. The other SmartBlock is a conventional grid block system without fastening surfaces (a single, all-foam unit).

Energrid Inc.

1-13
An ENER-GRID panel.

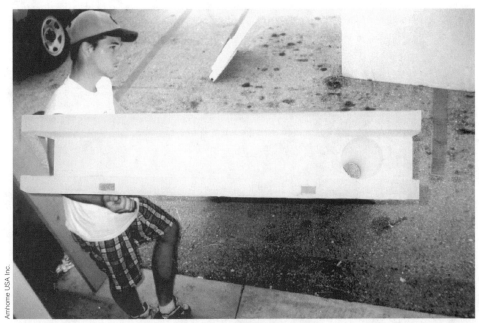

Amhome USA Inc.

1-14 *Top view of an Amhome panel with embedded furring strips.*

1-15
*Folding out a Fold Form
block before use.*

American ConForm Industries, Inc.

1-16 *The two varieties of SmartBlock: a flat block assembled with plastic ties (far left, third and fourth from left) and a grid block without fastening surfaces (second from left).*

2

Tools and materials

Some of the tools and materials commonly used in ICF construction are unfamiliar to many frame builders, and some familiar tools and materials are used in unfamiliar ways. Table 2-1 summarizes the products that have worked well.

Cutting and shaving foam

You need to cut the foam units to make holes for utility lines, and cut or shave them to make up correct lengths and heights at the edges of the walls and openings. If you are using a system with steel ties, you need to pick tools that cut metal as well as foam.

Carry a drywall or keyhole saw for making holes and curved cuts. For small, straight cuts and shaving the edges of a unit, a fine-toothed handsaw like a PVC or mitre saw is handy yet precise. Shaving can also be done with coarse (50-grit or flooring-grade) sandpaper or a hand rasp.

To make cuts completely through flat panel and the flat block units that are assembled in the field, most of the cuts are made on single sheets of foam before assembly. Any one of the handsaws mentioned previously will do the job well. For very fast cutting, try something with coarse teeth like a bow saw or garden pruner. For cutting through units of the other systems (which are pre-assembled), you can use an ordinary circular saw. If your system has steel ties, turn the blade backwards or use a metal-cutting blade. ICF units are thick, so you will have to cut twice to go completely through, once on each side. You can line up the two cuts precisely with a square, but usually, eyeballing carefully is precise enough.

You can also use a reciprocating saw to cut completely through a unit. With a long blade you can do it in one pass. This saw is especially well suited to making cuts in place, such as taking a small opening out of an erected foam wall. However, it requires a steady hand and cuts less straight than a circular saw.

The thermal cutter is a new tool that cuts a near-perfect line through foam and plastic units in one pass. Figure 2-1 contains a picture. A taut wire mounted on a bench is heated with electricity and drawn through the unit. It melts a nar-

Table 2-1 Useful Tools and Materials

Operation or class of material

Tool or Material	Comments
Cutting and shaving foam	
Drywall or keyhole saw	For small cuts, holes, and curved cuts.
PVC or mitre saw	For small, straight cuts and shaving edges.
Coarse sandpaper or rasp	For shaving edges.
Bow saw or garden pruner	For faster straight cuts.
Circular saw	For fast, precise straight cuts. For cutting units with steel ties, reverse the blade or use a metal-cutting blade.
Reciprocating saw	For fast cuts, especially in place.
Thermal cutter	For fast, very precise cuts on a bench. Not suitable for steel ties or grid panel units.
Chain saw	For fast cuts of grid panel units.
Lifting units	
Forklift, manual lift, or boom or crane truck	For carrying large grid panel units and setting them in place. For upper stories, a truck is necessary.
Gluing and tying units	
Wood glue, construction adhesive or adhesive foam	
Small-gage wire	For connecting units of flat panel systems.
Bending, cutting, and wiring rebar	
Cutter-bender	
Small-gage wire or precut tie wire or wire spool	
Filling and sealing formwork	
Adhesive foam	
Placing concrete	
Chute	For below-grade pours.
Line pump	Use a 2-inch hose.
Boom pump	Use two "S" couplings and reduce the hose down to a 2-inch diameter.
Evening concrete	
Mason's trowel	
Dampproofing walls below grade	
Nonsolvent-based dampproofer or nonheat-sealed membrane product	
Surface cutting foam	
Utility knife or router or hot knife	Heavier utility knives work better. Use a router with a half-inch drive for deep cutting.
Fastening to the wall	
Galvanized nails, ringed nails, and drywall screws	For attaching items to fastening surfaces. Use screws only for steel fastening surfaces.
Adhesives	For light and medium connections to foam.

Operation or class of material

Tool or Material	Comments
Insulation nails and screws	For holding lumber inside formwork.
J-bolt or steel strap	For heavy structural connections.
Duplex nails	For medium connections to lumber.
Small-gage wire	For connecting to steel mesh for stucco.
Concrete nails or screw anchors	For medium connections to lumber after the pour.
Flattening foam	
Coarse sandpaper or rasp	For removing small high spots.
Thermal cutter	For removing large bulges.
Foam	
Expanded polystyrene or extruded polystyrene	Consider foam with insect-repellent additives
Concrete	
Midrange plasticizer or superplasticizer	For increasing the flow of concrete without decreasing its strength. Can also be accomplished by changing proportions of the other ingredients.
Stucco	
Portland cement stucco or polymer-based stucco	

row path through foam and plastic ties. You might find the investment worthwhile if you are building a high volume of ICF walls. It will not cut through metal ties or the foam-and-cement material of the grid panel systems, however. Companies selling thermal cutters are listed in the appendix of this book.

Although the grid panel systems can be cut with any of the bladed tools previously mentioned, a chain saw is sometimes handy for them. It goes quickly through the heavier material of these systems and cuts through in one pass, whether cutting on the ground or in place.

Lifting units

Even the largest units of most ICF systems are light enough for one person to handle. Since the grid panel systems are made of a heavier material, they require two people to lift and carry the smaller (15-inch-wide) units and lifting equipment for the larger (30-inch-wide) ones. Any of the following can do the job: fork lift, manual lift, boom truck, or crane truck. The forks of these machines fit neatly into the panel's cavities. If you build more than one story, you will need one of the trucks, which have a higher range.

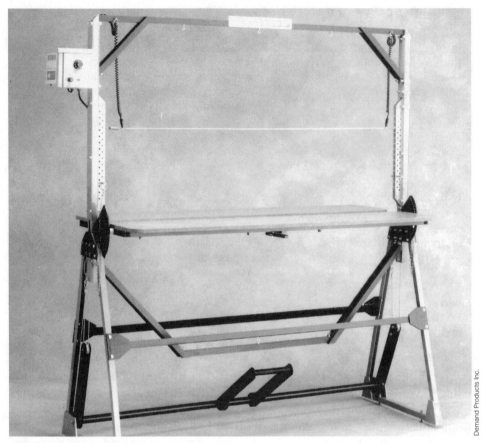

Demand Products Inc.

2-1 *Thermal cutter.*

Gluing and tying units

ICF units are frequently glued at the joints to hold them down, hold them to-gether, and prevent concrete leakage. Common wood glue and most construc-tion adhesives do the job well. Popular brands are Liquid Nail, PL200, and PL400. Some of these can dissolve foam, but if applied in a thin layer, the amount of foam lost is usually insignificant. To be careful, look for an adhesive that is "compatible with polystyrene."

Adhesive foams are a popular alternative. They expand to fill gaps and maintain insulation, although they are more expensive. Adhesive foams are widely available in pressurized cans that work like a can of whipped cream. Great Stuff is a well-known brand. But for an entire house, industrial products with applicator guns (as pictured in Fig. 2-2) are more practical. The gun starts and stops more immediately, industrial foam cures quickly, and the cans hold more. Industrial guns and foam are available from foam supply houses.

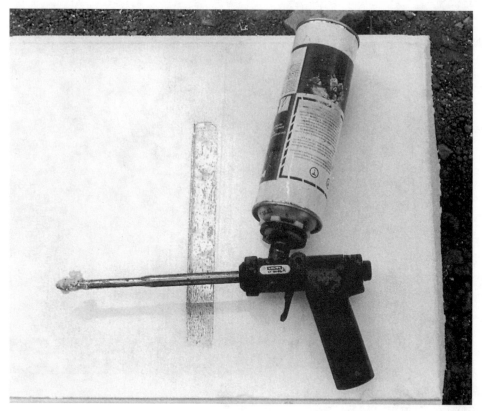

2-2 *Industrial foam gun.*

The flat-panel systems often use wire instead of glue to tie adjacent units together. Almost any small-gage wire will do, including a spool of rebar wire.

Bending, cutting, and wiring rebar

Rebar is often precut to length and prebent, but even if it is, the workers generally have to process a few bars in the field. Most ICF systems also have cradles that hold the bars in place for the pour, but a few bars need to be wired to one another or to ties to keep them in the position.

It is possible to bend rebar with whatever tools are handy and cut them with a hacksaw. But if you do large quantities, you might prefer buying or renting a cutter-bender, a large manual tool pictured in Fig. 2-3 that makes the job faster and easier. Cutter-benders are available at steel-supply, concrete-supply, and masonry-supply houses.

Almost any steel wire can hold rebar in place. But most efficient are rolls of precut tie wires (pictured in Fig. 2-4) or wire coils and belt-mounted coil holders, both of which are sold by suppliers of concrete products, masonry, and steel.

2-3 *Rebar cutter-bender.*

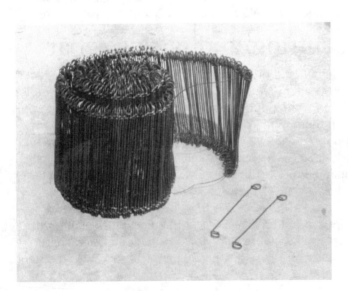

2-4
Roll of precut tie wires.

Filling and sealing formwork

The foam surface of the ICF formwork can get a variety of nicks and surface cuts. Any of these that penetrate or greatly weaken the formwork must be sealed before pouring to prevent leaks and blowouts. It is also a good idea to fill in any remaining large nicks and cuts before the foam is covered to maintain the insulation and form a consistent backing for the finish materials. Do filling and sealing with one of the adhesive foams mentioned previously.

Placing concrete

Concrete is best placed at a more controlled rate into ICFs than it is into conventional forms. An ordinary chute can be used for foundation walls (basement or stem), and this is the least expensive option because it comes free with the concrete truck. But precise control is more difficult with a chute. You must place more slowly than with conventional forms and you must move the chute and truck frequently to avoid overloading any one section of the formwork.

Most used above grade (and sometimes used below) are the line pump (also called a grout pump) and boom pump. Base which you use on cost and availability of a suitable rig. Both can be rented from equipment companies, and some contractors have their own.

The smaller line pump pushes concrete through a hose that lies on the ground. Figure 2-5 contains a photo. The crew holds the end of the hose over

2-5 *Line pump.*

the formwork to drop concrete inside. If possible, use a 2-inch hose. One or two workers can handle it, and it can generally be run at full speed without danger. If only a 3-inch hose is available, you can use it, but pump slowly until you learn how much pressure the forms can take.

Boom pumps are mounted on a truck that also holds a pneumatically operated arm (the boom). The hose from the pump runs along the length of the boom and then hangs loose from the end (see Fig. 2-6). By moving the boom, the truck's operator can dangle the hose wherever the crew calls for it. One worker holds the free end to position it over the formwork cavities. The standard hose diameter is 4 inches.

2-6 *Boom pump.*

You will need to have the hose diameter reduced to 2 or 3 inches with tapered steel tubes called "reducers." The narrower diameter slows down and smooths out the flow of the concrete. Figure 2-7 shows how the reducers are arranged on a boom with hoses. Also ask for two 90-degree elbow fittings on the end of the hose assembly, as in Fig. 2-7. These form an "S" in the line that further breaks the fall of the concrete.

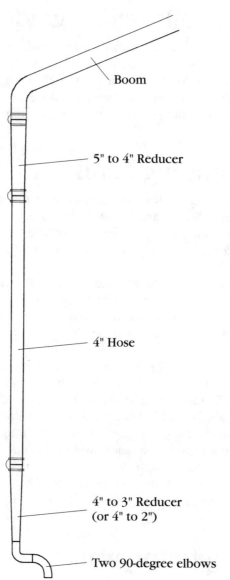

Boom

5" to 4" Reducer

2-7
*Diagram of boom pump
fittings suitable for pouring
ICF walls.*

4" Hose

4" to 3" Reducer
(or 4" to 2")

Two 90-degree elbows

Evening concrete

It is sometimes necessary to even out the concrete along the top of the form wall
after it is poured. You can do this with almost any straight edge, but a mason's
trowel is most efficient.

Dampproofing walls below grade

Basement walls made of ICFs should be coated with dampproofing or water-proofing material just as block and conventional poured walls are. Make sure the coating you use is labeled "solvent-free" or "nonsolvent-based," since materials containing solvents will dissolve the foam. Membrane products also work well, except for those that require heat sealing (usually with a blowtorch) at the seams. The heat melts the foam.

Surface cutting foam

It is necessary to cut narrow chases and shallow rectangles out of the interior surface of the foam for electrical lines, boxes, and an occasional pipe. The simplest tool for the job is a utility knife, the heavier the better. However, it is difficult to get a consistent cut with a knife, and the cutting produces a lot of small scrap.

A better alternative is an ordinary router. Router cutting is fast, the chase is consistent, and with the right bit the chase can be sized however you want. Cutouts for boxes are made with a few passes across the foam. The router, however, also produces unusable scrap. Use a router with a ½-inch drive to assure you can cut deep enough.

A new tool is the hot knife, pictured in Fig. 2-8. The hot knife is a corded, thermal tool like the thermal cutter. Electricity passes through the thin blade, heating it. You draw it through the foam for clean cuts about as fast as a router. U-shaped blades are available, as are bendable blades that you can form into any shape you want. When cutting chases, the cutout foam is one long piece that can be put back into the chase for insulation before wallboarding.

Note that none of these tools will cut through steel ties, so you need to cut around them. The hot knife will also not cut through the cement-foam mixture of the grid panel systems, although the other tools will. All will cut through plastic ties.

Fastening to the wall

Roofs, floor decks, interior walls, electrical boxes, trim, and fixtures are generally attached to the exterior walls. Some use ordinary fasteners and adhesives to connect to the foam units themselves, but in certain instances you will use special fasteners (pictured in Fig. 2-9) to attach to the concrete.

If your system has fastening surfaces, a lot of connections are done with certain wood nails or screws attached to them. On plastic ties you can use ringed or hot-dipped galvanized nails. Ordinary common nails pull out too easily. For screwing, use drywall screws. If your system has steel ties, use drywall screws only. Nails of any type do not grip well in steel.

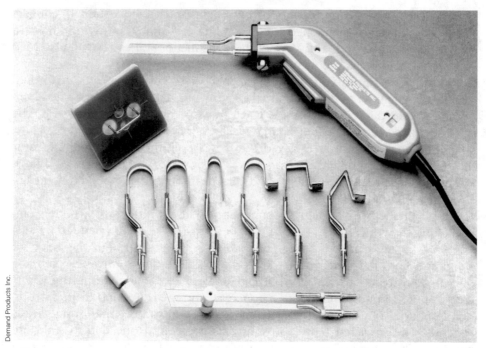

Demand Products Inc.

2-8 *Hot knife.*

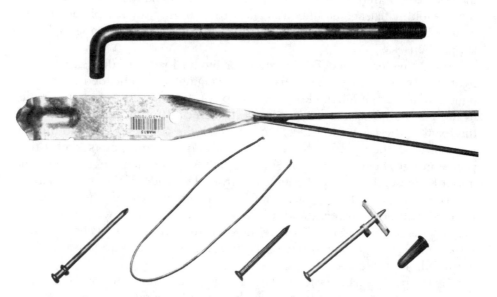

2-9 *Concrete fasteners (top to bottom and left to right): J-bolt, sill strap, duplex nail, light-gage wire shaped into a "U," concrete nail, powder-actuated pin, screw anchor.*

You can also glue items to the foam for light- and medium-strength connections. The adhesives mentioned previously under "Gluing and tying units" all hold wood securely to foam. Most of the construction adhesives will also hold solid plastic and metal products.

In a few situations you will need to nail or screw through the foam into pieces of 2-x lumber to hold them to the formwork. This need arises when lumber must be held in place inside the ICF formwork, making gluing difficult. You can use a large-headed nail or a large washer on a conventional nail or screw. Simple fasteners with ordinary-sized heads will pull through the foam. Ideal is the "insulation nail" or "insulation screw." It consists of a fastener with a large plastic washer preinstalled. They are available through many building- and insulation-supply houses.

Make the heaviest structural connections with either of two fasteners embedded in the concrete: a J-bolt or a steel strap. These are embedded by suspending one end in the cavity and keeping it there through the pour, or jamming it into the wet concrete right after. This requires advance planning on where connections will go, but it makes for extremely strong ones. The crook of the J-bolt is embedded in concrete, and the other end is threaded to bolt lumber or other components to the wall. Steel straps come in numerous varieties. The so-called hurricane strap has a crook at one end to sink into concrete. The other end is wrapped around and nailed to lumber, usually roof rafters or trusses. The so-called sill strap has two free ends that stick out of the concrete, both of which are wrapped around and nailed to a piece of lumber to hold it flat against the wall. It is used mostly for sill plates and top plates. Both products are available through concrete-supply and building-supply houses.

Some smaller embedded fasteners for lighter loads are duplex nails and light-gage wire. Duplex nails (also called staging or two-headed nails) hold light and medium lumber, like furring strips or the end studs of interior walls, to the ICF wall face. Place them inside the erected forms and push through the foam so the point protrudes. Once placed, the concrete will hold the nail by its heads. Then pound the wood over the protruding points, and if they poke through, bend them over. For wire, cut 8- to 12-inch lengths and bend into a U shape, then poke the ends out from the inside of the forms, leaving the rounded end of the U protruding back an inch or so inside the cavity. The concrete will hold this end, and the two loose ends can be wrapped around whatever is to be fastened. This technique is used mostly on walls that need a steel mesh across their surface for stuccoing: the wires are twisted to the mesh to hold it in place.

There are also many fasteners to put into the concrete after pouring. There is now a wide variety of concrete nails, including loose ones driven with an ordinary hammer and so-called "powder-actuated pins" (also called "Ramsets" after the most common brand). The pins are loaded into a special gun, along with a small shell of powder that shoots them like a bullet. Some pneumatic nail guns can also drive a concrete nail, with the nails available in a strip like wood nails. All of these are sold in building supply stores.

You can also use ordinary wood screws with plastic anchors. This requires two steps: drilling with a concrete bit, then inserting the anchor and turning the screw.

Flattening foam

After the concrete is placed, occasionally you will cut or shave the surface of the foam to flatten out an occasional high spot. The foam can be shaved with a coarse rasp or a large sanding block and heavy-grit sandpaper. There is also a special thermal tool, called a thermal skinner, designed to shave large bulges quickly. Figure 2-10 shows a picture of one. The taut wire of the skinner is electrically heated to slice through the foam.

Foam

Although with most systems the foam is provided, if you are using a grid panel, flat panel, or flat plank system, you usually buy your own foam. And some of the other systems give you some choice, as well.

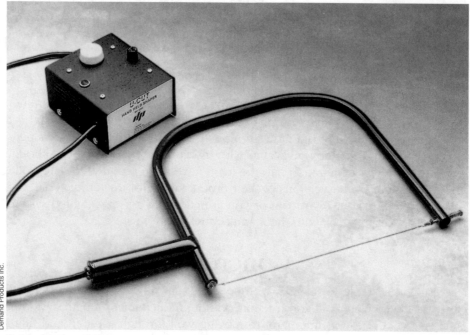

Demand Products Inc.

2-10 *Thermal skinner.*

By far the most common foams are expanded polystyrene (often abbreviated EPS) and extruded polystyrene (XPS). EPS consists of tightly fused beads of foam. Vending-machine coffee cups, for example, are made of EPS. XPS, produced in a different process, is more continuous, without beads or the sort of "grain" of EPS. The trays in prepackaged meat at the grocery store are XPS. Figure 2-11 pictures both types. The two types can differ in cost, strength, R-value, and water resistance. If you are buying your own foam, check the price and specifications of the foam available in your area.

2-11 *Foams: expanded polystyrene (EPS, left) and extruded polystyrene (XPS).*

The EPS varies somewhat. It comes in various densities, the most common being 1.5 pounds per cubic feet (pcf) and 2 pounds pcf. The denser foam is a little more expensive, but is a little stronger and has a slightly higher R-value.

Some stock EPS is now available with insect-repellent additives. Although few cases of insect penetration into the foam have been reported to date, some ICF manufacturers offer versions of their product made of treated material, too. Whether you buy your own foam or you are choosing a preassembled system, you might want to check with the manufacturer about this.

Concrete

It is usually desirable to fill ICFs with concrete that flows better than the concrete used in conventional construction, without reducing its strength or adding great cost. Good flow is important to moving well through a pump and filling the formwork cavities thoroughly. You improve flow by varying the proportions of the ingredients or by using a plasticizer.

Ordinary concrete is a mixture of portland cement, water, sand, and gravel or crushed stone. Supplementary cementing materials such as fly ash are also sometimes added. The gravel or crushed stone is referred to as coarse aggregate and the sand as fine aggregate. The water and cementing materials form a paste that gradually hardens, binding the aggregate into a solid concrete mass. Concrete stiffens noticeably within a few hours of mixing, feels solid after a day, and approaches its maximum strength in a month or so.

The ease with which concrete flows depends primarily on the proportions of ingredients used and on the maximum size of the coarse aggregate. There are four common ways to improve flow, some of which can be used in combination. One is to use more water. This adds no cost but lowers the strength and makes the hardened concrete less watertight. It is therefore not recommended. A better method is to use more cement and less aggregate. This adds some cost but doesn't sacrifice strength or watertightness. A third way is to limit the size of coarse aggregate to small pieces. This also adds some cost because the amount of sand, water, and cement must be increased. The fourth way is to add a chemical called a "plasticizer" that improves flow without lowering strength. It comes in different varieties, including superplasticizers, which have a big effect, and midrange or normal plasticizers, which have a smaller effect on flow.

Chapter 6 gives instructions on what concrete mix to use under what circumstances. However, you will need to know how concrete with specific properties is commonly ordered. The purchaser usually tells the concrete supplier the compressive strength needed, the maximum aggregate size, and the slump when the concrete is first mixed. Compressive strength is given in pounds per square inch (psi), which is the maximum pressure the concrete can withstand. Maximum aggregate size sets a limit on the largest piece that's permitted. In concrete with a "¾-inch" aggregate, no piece is supposed to measure more than ¾ of an inch across. Slump is an indicator of the concrete's ability to flow. It's measured by filling a 12-inch-high tapered mold with concrete, using a rod to assure proper compaction. After the mold is slowly removed, the concrete sags or slumps. If it drops down an inch (forming an 11-inch-high pile) the mix is said to have a "1-inch slump;" if the drop is 2 inches, the mix has a 2-inch slump, and so on.

You should also tell the concrete supplier if the concrete is to be pumped. Concrete to be pumped through a small-diameter line will usually be made with a smaller aggregate, more cement, and more sand than is used in normal concrete. Some experienced contractors specify the concrete ingredients and proportions (gallons of water, pounds of cement, and so on) themselves to achieve the combination of flow and economy that they need. When doing this, they assume responsibility for the strength and other properties of the concrete. See chapter 6 for more information.

Stucco

ICFs can take any conventional siding material, but when it is stucco you should decide the type you will use early on. Cement stucco (also called Portland PC or traditional stucco) normally requires a steel mesh placed over the foam. As noted later, you sometimes need to install the fastener to attach the mesh while setting the formwork. Polymer-based (or PB) stuccos do not normally use steel mesh. Instead, the crew embeds a fiberglass mesh into the stucco while applying it. In this case you install no fasteners while setting the forms.

3

Design and planning

By taking the properties of your ICF system into account from the beginning of your project, you can improve your efficiency of construction, cut costs, and improve the quality of the final structure. Table 3-1 contains key data that is important to design and planning. Note, however, the single most important rule of ICF design: Consult the manufacturer's technical documents, the manufacturer directly, or an engineer for all structural design decisions.

Picking a width

As noted in Table 3-1, the units of several systems come in different widths. The different widths form concrete walls of different thicknesses to meet different structural requirements. If you use one of these systems, consult the technical documents, the manufacturer, or an engineer to determine which width to use. However, for initial planning purposes you can assume that you will need a unit of a width (unit "total width" in Table 3-1) that forms a concrete wall about 6 inches thick (concrete "maximum thickness"). For some full basements, you will need to use the next widest unit to withstand the earth, groundwater, and loading forces.

House design

Virtually any house design that can be built with wood or steel frame can be built with ICFs. However, an ICF wall has a greater thickness, weight, and R-value, and this makes it useful to make a few adjustments in the plans.

Overlay on existing plans

In many cases you will begin with a set of house plans designed for frame. The greater thickness of the ICF walls requires you to choose whether to line up the interior surface of the ICF walls with the interior of the walls on the plans, or line up the exteriors.

It usually requires less change to line up the interiors. It takes no space from inside the house and requires no adjustment to the interior fixtures, walls, or dimensions. But remember to shift the footer or foundation out so that it is cen-

Table 3-1. Important ICF Design and Planning Data

	Unit dimensions			Concrete		
	Width	Area	Perimeter	Maximum Thickness	Yards Per unit	Research Reports
Panel systems						
Flat panel systems						
Styro-Form	10"	16 sf	20'	6"	.3	None /a
	12"	16 sf	20'	8"	.4	
	14"	16 sf	20'	10"	.5	
	16"	16 sf	20'	12"	.6	
R-FORMS	8"	32 sf	24'	4"	.4	None /a
	10"	32 sf	24'	6"	.6	
	12"	32 sf	24'	8"	.8	
Grid panel systems						
ENER-GRID /b	8"	12.5 sf	22.5'	4"	.042	None
	10"	12.5 sf	22.5'	6"	.12	
	12"	12.5 sf	22.5'	6"	.12	
RASTRA /b	8"	12.5 sf	22.5'	5.9"	.106	ICBO 4203
	10"	12.5 sf	22.5'	6.3"	.12	
	12"	12.5 sf	22.5'	6.3"	.12	
	15"	12.5 sf	22.5'	6.3"	.12	
Post-and-beam panel systems						
Amhome	9⅜"	32 sf	24'	5½"	.13/a	None
Plank systems						
Flat plank systems						
Diamond Snap-Form	8"	8 sf	18'	4"	.1	None /a
	10"	8 sf	18'	6"	.15	
	12"	8 sf	18'	8"	.2	
	14"	8 sf	18'	10"	.25	

Lite-Form	12"	5⅓ sf	17'4"	8"	.133	None /a
Polycrete	11"	8 sf	18'	6"	.15	Canadian /a
	13"	8 sf	18'	8"	.2	
	15"	8 sf	18'	10"	.25	
QUAD-LOCK	8"	4 sf	10'	3½"	.05	ICBO 5188 /a
	11"	4 sf	10'	5½"	.07	
	13"	4 sf	10'	7½"	.10	
	15"	4 sf	10'	9½"	.12	

Block systems

Flat block systems

AAB	11½"	5.58 sf	108"	6¼"	.111	BOCA 94-31, SBCCI 94172, ICBO 5119 /a
	12½"	5.58 sf	108"	8"	.138	
Fold-Form	8"	4 sf	10'	4"	.049	None /a
	10"	4 sf	10'	6"	.074	
	12"	4 sf	10'	8"	.0999	
GREENBLOCK	10"	2.78 sf	5'10"	5⅓"	.46	None
SMARTBlock Variable Width Form	8"	3⅓ sf	8'8"	3¾"	.04	BOCA 89-67, SBCCI 9426, ICBO 4572 /a
	10"	3⅓ sf	8'8"	5¾"	.06	
	12"	3⅓ sf	8'8"	7¾"	.08	

Grid block systems with fastening surfaces

I.C.E. Block	9¼"	5⅓ sf	108"	6⅜"	.075	BOCA 92-28, SBCCI 94177, ICBO 5055
	11"	5⅓ sf	108"	8"	.1	
Polysteel	9¼"	5⅓ sf	108"	6⅜"	.074	BOCA 90-16, SBCCI 9230, ICBO 4295
	11"	5⅓ sf	108"	8"	.1	
Reddi-Form REWARD	9¼"	5⅓ sf	108"	6⅜"	.074	None
	11"	5⅓ sf	108"	8"	.1	

Table 3-1. Continued.

	Unit dimensions			Concrete		Research Reports
	Width	Area	Perimeter	Maximum Thickness	Yards Per unit	
Therm-O-Wall	9¼"	5⅓ sf	10'8"	6⅜"	.074	None
	11"	5⅓ sf	10'8"	8"	.1	
Grid block systems without fastening surfaces						
Reddi-Form	9⅝"	4 sf	10'	6½"	.05	BOCA 90-81
SmartBlock Standard Form	10"	2.78 sf	8'4"	6½"	.05	BOCA 89-67, SBCCI 9426, ICBO 4572
Post-and-beam block systems						
Energy Lock	8"	1.78 sf	6'8"	5"	.018/c	None
Featherlite	8"	⅞ sf	4'	5"	.006/c	ICBO4643
KEEVA	8"	4 sf	10'	5"	.018/c	None

/a Because of flat systems' similarities to conventional concrete forms, building departments tend to request research reports for them less often.

/b The grid panels are available in certain heights and lengths other than the standard 15" height and 10' length. The areas, perimeters, and yards of concrete per unit of these other-dimensioned panels will differ accordingly.

/c Amounts of concrete per unit vary with post-and-beam systems. With these systems, only some cavities are filled with concrete. The figures given here are averages for planning purposes only. Consult the manufacturer to calculate the amount of concrete per unit that results from your particular fill pattern.

tered under the ICF walls. Also remember that your roof will be slightly larger. And recheck your setback requirements. If they were tight before, the thicker ICF wall could put your house over the allowable line.

Table 3-1 includes the width of all ICF units ("total width") to use in calculating new plan dimensions. This width is the distance between the inside and outside faces of foam, which is the thickness of the final wall without finishes.

Projections and load-bearing spans

Although they have been built, ICF walls that do not line up with the foundation require extra consideration because of the weight of the concrete that is unsupported above. You can handle features like these by taking them out altogether (as depicted in Fig. 3-1) or leaving them in and having an engineer design their details.

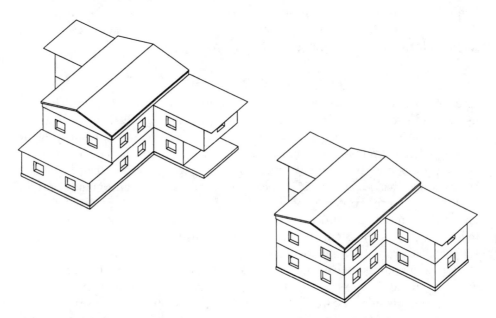

3-1 *House with a load-bearing span and a projection (left), and a redesigned version that eliminates them (right).*

Projections are sections of wall that stick out farther than the walls below them (as shown in Fig. 3-1). Typical examples include garrisons and some bays. In frame construction they are usually supported by a floor that is cantilevered to project out a couple of feet. If you prefer to keep a projection in the design, you can build it of ICFs if the cantilevered floor below is designed to carry the extra weight of the concrete. This requires engineering. Or you can cantilever as you normally would and build the walls of the projection out of frame.

Load-bearing spans are walls that are set back from the walls below (as depicted in Fig. 3-1) and therefore are also unsupported. Again, you can line up the walls to remove the load-bearing span, or you can build it similarly to the way it would be built with frame: build the floor deck below to support the weight, add structural posts or columns inside the house to support the wall above, or add sufficient strength to the span itself so that it acts as a beam. If the span wall is short it can usually be strengthened sufficiently by adding extra rebar at the bottom. But for a long span, a steel or wood beam might be necessary. In any case, an engineer should design the structural components of the wall and the floor or posts below.

Other shape considerations

The Amhome system includes not only ICF walls, but also a roof system constructed out of wooden I-joists and EPS foam for superinsulation on top. It is helpful when using this system to use a gable roof and avoid using many corners in the walls. Complex roof styles (hips, gambrels) and frequent corners will require time-consuming cutting of foam during roof construction.

Opening placement

As a rule of thumb, there should be at least 16 inches of wall above and along either side of every door and window, as shown in Fig. 3-2. This means, first, that there will be at least 16 inches of ICF wall between each pair of openings that are side by side. That provides enough room for concrete to flow into the cavities and enough strength in the column between the openings to support the wall above. Second, there will be at least 16 inches of ICF wall between an opening and the next opening or the roof line above it. That provides room to form a lintel strong enough to support the structure above. For wide openings, more room above might be necessary. For more exact requirements in all of these matters, consult the ICF manufacturer's technical documents, the manufacturer directly, or an engineer.

Openings can be placed closer together by using the arrangement shown in Fig. 3-2. Frame the portion of the wall between the two openings instead of putting concrete in it, and treat it as one large opening when doing structural design. In other words, when deciding how much wall and rebar to put above the opening, assume it has a width equal to the total widths of the two separate openings plus the framed wall in between.

Non-ICF walls

You might prefer to build some of your walls out of conventionally formed concrete or frame. If the foundation includes stem walls or a basement and there are no plans to insulate or finish them, these walls can usually be built less expensively with concrete block or concrete poured into conventional forms. In addition, post-and-beam panel systems are not appropriate for below-grade walls

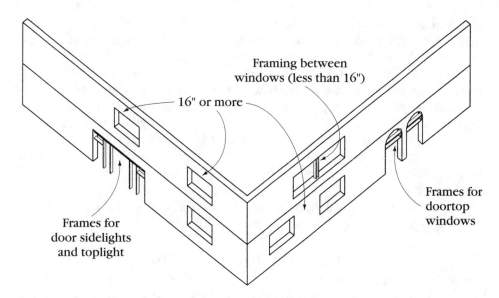

3-2 *Openings spaced or framed to meet the requirements of ICF formwork.*

because their long expanses of unsupported foam are not designed to withstand backfill and groundwater forces. The post-and-beam block systems must be filled completely (every cavity) with concrete below grade.

If you do form the foundation out of conventional block or conventional poured concrete, make the thickness of the foundation's walls at least as great as the maximum concrete thickness of the ICF system you will use above it (see Table 3-1), and center the ICF wall on the foundation wall. This places all the concrete of the ICF wall squarely over the concrete of the foundation.

If the attic walls will not be insulated or finished, the ends of a gable or gambrel roof can sometimes be built less expensively out of frame. And projections can be simpler out of frame because of their lighter weight. Details on tying frame walls to ICF walls are in the appropriate sections later in the manual.

If the difference in cost and difficulty appears to be small, it is usually wiser to build odd walls with ICFs, for a few reasons. Switching back and forth between structural systems takes time. Frame walls are usually less strong and energy efficient. A conventional concrete basement serves as a channel for heat to flow out of the ICF wall, reducing its energy efficiency. So if you use a conventional basement, at least consider insulating it on the outside to prevent this.

Door and window depth

Position exterior doors that swing inward flush with the interior of the ICF wall. If they are flush with the exterior, the thick wall will make it impossible to swing the door open the full 180 degrees. Any exterior doors that will swing outward should, conversely, stand flush with the exterior surface of the wall. Note also that a door with a sidelight or a framed portion of wall to the hinge side is exempt from this rule; the door can swing flat to this thinner section of wall.

Windows can be set at any depth in the opening. Conventional (flanged) frame windows are mounted flush with the exterior face, as they are normally. This will leave a deep window box on the interior, including a deep sill that can serve as a shelf or narrow window seat. Flangeless (also called in-wall mount or screw-through) windows can be mounted at any depth. If you choose to mount them back from the exterior face, you will have a recessed look outside, an exterior window box to finish somehow, and less depth to the interior window box.

Remember that you will need to order doors with extended jambs and wider window box trim (if any) to span the greater thickness of the ICF wall.

Rebar and concrete pattern

For ICFs, the principle means of increasing the strength of the wall is adding greater amounts of steel reinforcing bar. With post-and-beam systems, you will also add more concrete; since the wall is filled only in those cavities that contain rebar, whenever rebar is added to a new spot, more concrete is used as well to fill the surrounding cavity. Before building it is necessary to determine how much rebar must go into what positions on each wall. As with any structural decision, consult the ICF manufacturer's technical literature, the manufacturer directly, or an engineer to learn the correct pattern of rebar and concrete to use. The following guidelines are intended for rough planning purposes only.

For below-grade walls, the pattern of rebar will usually be close to what you use in a conventional poured concrete below-grade wall of the same concrete thickness. For above-grade walls, correct placements of concrete and rebar depend on load conditions (the height of the wall, whether there are other stories above, local wind and seismic hazards, and so on). For a one-story house with 8-foot walls in an area of moderate wind and seismic hazards, a typical ICF rebar pattern is depicted in Fig. 3-3. One #4 (⅝-inch diameter) bar stands vertically every 24 inches on center (oc), with one additional vertical #4 bar at each corner and along each side of every opening. There is also a horizontal #4 bar every 32 inches oc. Over narrow (less than 6 feet wide) openings run two additional #4 rebar to form a lintel. Wider openings require more bars, larger bars, or some combination of these. Vertical bars from the floor below and the floor above overlap by at least 2 feet, as do horizontal bars. The extra bars alongside or above an opening run beyond it at least 2 feet in each direction.

In greater load conditions, the wall is strengthened by using a larger size of bar, more bars, or both. In a post-and-beam system, there is usually less rebar for above-grade walls. A typical pattern for post-and-beam block systems in moderate load conditions includes the same vertical bars listed previously, but a horizontal bar only every 4 or 8 feet. There are also no extra bars over lintels;

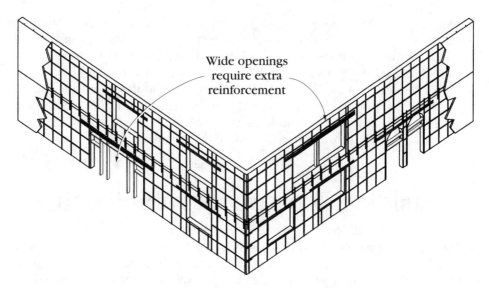

Wide openings require extra reinforcement

3-3 *A typical pattern of steel reinforcing bars in an ICF wall.*

the vertical bars along either side of the opening go all the way to the next scheduled horizontal bars above. For the post-and-beam panel (Amhome) system, the pattern is similar, except that #5 bars are used and the regularly spaced vertical bars come every 4 feet, not every 24 inches.

Whatever system you use, there is a useful general rule: when in doubt use more reinforcing. Enlarge the size of the bars you are using or use more frequent bars. Steel is relatively inexpensive, so the precaution costs little. The only serious potential problem from oversizing rebar is that it will be too close together or packed too tightly into the smaller cavities of the formwork. This can cause difficulties during the pour by preventing the concrete from flowing. It can also prevent sections of bars from being adequately surrounded by concrete, decreasing total wall strength. So plan to leave space between bars and avoid clogging narrow sections of the cavities.

Utility line placement

With most ICF systems, it is useful to know where you will need penetrations through the exterior walls (such as for gas service, water service, electrical and phone service, sewage lines, outdoor spigots, piping to outdoor AC units, and so on) before pouring the concrete. You can then provide holes for them in the formwork, avoiding the need to drill concrete later. Consult with the trades that will need penetrations and keep a diagram that shows the desired size and location for each. Keep this with the plans. Note that this practice is usually unnecessary for post-and-beam systems above grade, since they have large sections of unfilled foam that can be cut easily later.

As with all building systems, it is also wise to avoid running large utility lines along exterior walls. This reduces insulation and, in extreme cases, can weaken the wall. Wiring and piping up to 1 inch in diameter can be cut into the foam surface of almost all ICFs without problem. However, avoid running vent stacks, ductwork, or drains in the exterior walls. If they must go there, consider building a frame chase along the wall to hold them so that they do not need to cut into the formwork. Failing that, note carefully where they will go on the plans so the wall crew can provide for them inside the formwork.

Building department submission

The materials you will have to submit to receive approval for an ICF house vary with the preferences of the local building department. Most departments will want, in addition to the floor plans, details that show the amount and placement of rebar (and, in the case of post-and-beam systems, amount and placement of concrete). Other departments that want more detail can be satisfied with the manufacturer's literature, which often includes engineering data and details. Call the manufacturer for free copies.

Even more particular departments might request a research report. A research report is a brief document written by one of the model code organizations in the United States (BOCA, SBCCI, or ICBO) or Canada (Canadian Construction Materials Center). The report provides engineering data on a system, describes proper design and installation, and lists the uses for which the system meets code. Many of the systems have a report from at least one of the code organizations. Table 3-1 lists the identification numbers of the reports currently available for each system. The manufacturers will usually provide copies free. The building department will prefer or require a report from the one code organization it has adopted. In most of the departments in the northeastern United States this is BOCA. In most of the Southeast it is SBCCI, and in most of the West ICBO. Note, however, that building departments rarely require a research report for any of the flat systems (flat panel, flat plank, flat block). Since the concrete forms a flat wall just as with conventional forms, the department is usually satisfied to treat it like a conventional poured wall.

If your building department requests a research report and none is available, the department will almost always settle for an engineer's review and stamp on the plans. In rare cases the department might ask for this even if there is a report. Almost any structural engineer is qualified to do the work. Many manufacturers will provide names of engineers who are experienced and efficient at it and licensed in your state.

Takeoff and cost estimation

The documentation for many systems includes rules or tables for doing precise materials takeoffs and job cost estimations. As you become experienced with a system you will probably develop your own tables. Following is a general procedure to help you understand the process and to use as a starting point when your system's documentation has no guide.

Table 3-2 lists the major items included in an ICF takeoff. The table also shows which of these items are necessary for which systems. Certain components are used in some systems but not others.

Table 3-3 is a worksheet for recording your takeoff. The sections below explain how to fill in each part. You might want to make several copies of the table rather than write on the original.

ICF units

The units of most systems are preassembled. The others involve some field assembly, so the takeoff must include each piece. To estimate the number of units necessary for any preassembled system:

1 Calculate the total exterior wall area to be constructed of ICFs. Leave out the area of large openings (anything bigger than 4 square feet).
2 Multiply the square-foot area you calculated by 10 percent (a waste factor) and add it to the original amount. Call this the "total square footage" and enter it into Table 3-3. Your waste should decline to under 5 percent in later houses, as you learn to use odd cuts of the foam units. If for any reason you are concerned you will run short, consider increasing the waste factor. Many manufacturers will take back and refund unused units.
3 If your system uses special units (for example, some have preformed or precut corners and post-and-beam block systems require "lintel block" everywhere a beam will go), count the number of these that you will need. Enter the name and quantity of each type of special unit into Table 3-3 for future reference. Calculate the exterior surface area of these units, multiply that by the number of units you will need, and subtract the resulting total area of special units from the total square footage you calculated in step 2. Call this the "standard square footage" and enter it into Table 3-3 as well. You will need to consult the manufacturer or the manufacturer's documentation to do some of these steps.
4 Find in Table 3-1 the unit's "wall area" of the system you are using.
5 Divide the standard square footage by the wall area to get the total number of standard units. Enter this into Table 3-3.

Table 3-2. ICF Construction Takeoff Items

Item	Type of System						
	Flat Panel	**Grid Panel**	**Post-&-Beam Panel**	**Flat Plank**	**Flat Block**	**Grid Block**	**Post-&-Beam Block**
ICF Units							
panels	X (Styroform)	X					
planks				X			
blocks					X	X	X
foam faces					X (SmartBlock)		
cavity plugs							X
corner faces					X (SmartBlock)		X (SmartBlock)
sheet foam	X (R-Forms)						
ties	X (R-Forms)		X	X	X (SmartBlock)		
fastening surface							
connecting channel	X (R-Forms)		X	X (Polycrete)			
Concrete	X	X	X	X	X	X	X
Steel reinforcing bar	X	X	X	X	X	X	X
Equipment rental							
concrete pump	X	X	X	X	X	X	X
lift		X					
Lumber							
bucks	X	X	X	X	X	X	X
bracing	X	X	X	X	X	X	X
Adhesive (glue units together)	X	X	X	X	X	X	X
Adhesive foam (fill and seal)	X	X	X	X	X	X	X
Fasteners	X	X	X	X	X	X	X
Labor	X	X	X	X	X	X	X

Table 3-3. ICF Construction Takeoff Form

Item

Total Square Footage | Square feet: _____

Special ICF units

   Type _____ | Number: _____
   Type _____ | Number: _____
   Type _____ | Number: _____

Standard Square Footage | Square feet: _____

ICF Units

   panels | Number: _____
   planks | Number: _____
   blocks | Number: _____
   foam faces | Number: _____
   cavity plugs | Number: _____
   corner faces | Number: _____
   sheet foam | Number: _____
   standard ties | Number: _____
   half ties | Number: _____
   corner faces | Number: _____
   fastening surface | Feet: _____

Concrete | Yards: _____

Steel reinforcing bar | Size | Length

		4'	8'	10'	Other:____	Other:____
	#3	____	____	____	_____	_____
	#4	____	____	____	_____	_____
	#5	____	____	____	_____	_____
	#6	____	____	____	_____	_____

Equipment rental

   concrete pump | Half days: _____
   lift | Days: _____

Lumber

   jambs and lintels | Dimensions: ____
   | KD or PT: _____ | Lineal feet: _____
   sills | (PT 2 × 2) | Lineal feet: _____
   flanges (¼-inch plywood) | Sheets: _____
   bracing 2 × 4 | Number: _____
         2 × 6 | Number: _____
      strapping | Bundles: _____

Adhesive (glue units) | Type: _____ | Tubes/cans: _____

Adhesive foam (fill and seal) | Cans: _____

Fasteners | Type: _____ | Number: _____
   | Type: _____ | Number: _____
   | Type: _____ | Number: _____
   | Type: _____ | Number: _____
   | Type: _____ | Number: _____

Labor | Crew-hours _____

For systems involving field assembly, calculate the number of units required as in steps 1–5. But before making entries for units into Table 3-3, break this down into the numbers of each component of the unit. You will need the manufacturer's information to do this. However, approximate numbers of components for each unit are:

- R-FORMS:
  ~2 sheets of 2"-x-4'-x-8' foam
  ~21 ties
  ~16–24 feet of connecting channel
- Amhome:
  ~1 sheet of 9¼"-x-4'-x-8' foam
  ~2 pieces of 8' strapping (as a fastening surface)
- Diamond Snap-Form:
  ~2 planks of foam
  ~8 ties, except at corners, special corner ties replace 3 standard ties
  ~8 half ties for each unit on very bottom or top course
- Lite-Form:
  ~2 planks of foam
  ~12 ties, except at corners, special corner ties replace 3 standard ties
  ~12 half ties for each unit on very bottom or top course
- Polycrete:
  ~2 planks of foam
  ~12 ties
  ~8 feet of fastening surface (plastic channel)
- QUADLOCK
  ~2 planks of foam
  ~5 ties
  ~5 half ties for each unit on very bottom or top course
- SmartBlock variable-width form
  ~2 foam faces
  ~2 corner faces (for units at a corner only)
  ~4 ties

Concrete

To estimate the amount of concrete necessary to fill the formwork:

1 Find in Table 3-1 the concrete "cubic yards per unit" for the system you are using. This is the yards of concrete needed to fill one unit.
2 Multiply this number by the number of units you previously estimated you will need, including both special and standard units.

Note that the cubic yards per unit will vary with post-and-beam systems. The stronger the wall needs to be, the more cavities will be filled with concrete and the greater the concrete per unit. Likewise, the concrete used goes up with the frequency of openings. The data in the table are for a typical above-grade wall. For more frequent reinforcing or openings, the cubic yards per unit will need to be increased proportionately. Consult the manufacturer's literature or the manufacturer.

Rebar

The size and amount of rebar needed will vary widely with such load factors as the height of the walls, number of stories, local wind and seismic conditions, local soil, and the number and width of openings. To estimate accurately:

1 Sketch elevations of the foundation and exterior walls and draw in the location of each bar, as was done in Fig. 3-3. Use a different line or color to draw each bar size (#3, #4, #5, #6).
2 Count up the number of each length of each bar size and enter the totals into Table 3-3.

Equipment

Figure a half-day rental of a concrete pump for each story of formwork. Omit the basement or stem walls if you plan to pour them with a chute. Add one more half-day for tall gable or gambrel ends that will be built of ICFs. Enter the number of half-days into Table 3-3.

If you are using a grid panel system and you do not own lifting equipment (as described in chapter 2), plan to rent some for about 2–3 days per floor. The time will be longer with large houses and will decline as your wall crew gains experience and works faster. You might, of course, currently own or buy the equipment instead. Enter the total number of days into Table 3-3.

Lumber

Dimension lumber is necessary to build bucks (also called subframes) to go inside of openings and to brace the formwork until the concrete has been poured and hardens. Note that some local building departments will allow lumber that abuts concrete to be ordinary kiln-dried (KD). Some will not accept plain kiln-dried, but will accept KD that is wrapped with tar paper to keep it out of direct contact. In either of these cases, buy kiln-dried lumber for the jambs and lintels of bucks, and use pressure-treated (PT) lumber for the sills. If your department will not accept KD of any sort, you will need to use PT lumber for all parts.

To take off the buck materials:

1 Measure the dimensions (width and height) of each opening.
2 Add the lengths of all jambs and lintels to get their total length and all sills to get a separate total. Increase each of these totals by 10 percent for waste. Your waste should drop with experience.
3 Enter the total length of the jambs and lintels into Table 3-3.
4 Now determine the width of buck lumber for jambs and lintels. If you will not be siding with stucco, find in Table 3-1 the total width of the ICF units you are using. The two widths must match, so if your units are 9¼" thick you will need 2-x-10 lumber, and so on. Write the lumber

width in Table 3-3 for future reference. If you will be siding with stucco, use the maximum width of the concrete (also in Table 3-1) instead of the total width of the units.

5 Also write in Table 3-3 whether you will use kiln-dried or pressure-treated lumber for jambs and lintels.

6 Double the total sill length calculated in (2) above, and enter that number into Table 3-3. The number is doubled because sills are built out of a pair of 2-x-2s. They are also built of pressure-treated lumber.

7 If you will not be siding with stucco, add the total lintel, jamb, and sill length, double it, and buy enough ¼-inch plywood to make 6-inch wide strips that total this length. These will form flanges for the bucks.

Bracing lumber is normally reused for the roof and interior framing. Therefore it is not an additional cost. However, you will need to place an early order for enough lumber to do the job. For the block systems, you can make a rough estimate of the amount of bracing lumber as follows:

1 Buy one 2-x-4 for each 2 lineal feet of exterior wall. If the walls of each floor will be 8 feet, make all of the 2-x-4s 8 footers. If the walls of any story will be taller, make at least half of the 2-x-4s as long as the wall height of the tallest floor.

2 For each corner, buy one 2-x-6 and one-half sheet of plywood. The 2-x-6s should be as long as the walls are tall. They will be used to form corner braces.

3 For each large opening, get enough 2 × 4s to place one vertically inside the opening every 3 feet (to hold the lintel up), and one horizontally every 4 feet (to hold the jambs apart).

4 Buy one bundle of strapping, or two for a very large house.

Figure this bracing lumber once, not for each floor separately—the lumber will be reused on each floor.

For the other (nonblock) systems the amount of bracing is different, as follows:

- Styroform, grid panel systems, post-and-beam panel systems, and flat plank systems: Buy about half as much of each type of lumber and plywood.
- R-FORMS: Custom bracing systems are available. Contact the manufacturer for details.

Adhesive

Most systems require gluing at least some of the units to the foundation and one another. Estimate the amount needed as follows:

1 Find the perimeter of the units (the lineal distance around the face of one unit) of your system in Table 3-1.

2 Multiply this by the number of units calculated and entered in Table 3-3 to get the total perimeter of all units. Include any special units.

3 If you will use construction adhesive, divide the total perimeter calculated in step 2 by 100. This is the number of 28-ounce tubes you will need. If you will use industrial adhesive foam, divide it by 600. This is the number of industrial-size cans you will need. Enter the number of tubes or cans into Table 3-3.

Note that some manufacturers claim their units will "friction fit" to one another. If you do this, you will only need about ⅓ as much adhesive.

For filling and sealing holes and gaps, you will also need some adhesive foam. (Construction adhesive will not work here.) To estimate the number of industrial cans you will need, take the total square footage you entered earlier into the table and divide it by 1000. Enter the result into Table 3-3.

Fasteners

The fasteners required vary widely from project to project, but they usually have a total cost about the same as for frame walls of the same size. Some of them are more expensive than nails, but you will use fewer. List out the heavy structural fasteners (J-bolts, sill straps) you think you will need to make connections to floor decks, roof members, and load-bearing walls. Then plan on buying a few boxes of duplex nails, a couple hundred powder-actuated pins, or a few extra tubes or cans of adhesives to make other connections.

Labor

Total labor will drop sharply with experience. For your first house, a reasonable estimate is that to erect the walls, a three-person crew headed by a skilled carpenter will take one minute for each square foot of wall. So:

1 Take the total square footage from Table 3-3.
2 Divide by 60 to get the number of crew-hours required to build the walls. Enter the result back into the table.

Note that complex wall features (bays, curved walls, and so on) will add to labor time.

When you estimate total costs, remember that the crew hours must be multiplied by the total hourly pay of all crew members.

Cost estimation

Once all materials, equipment, and labor are taken off and entered into Table 3-3, you can produce a total cost estimate. Get price quotes on all items in the table, multiply the amount of each item by its price to get the total cost for each item, and total these costs to get the estimated total project cost.

4

Project and site preparation

There are several steps to construction that must be done a little differently, even before the first ICF unit is set.

Picking crews

Take some care and thought when picking the crew to build the ICF walls. Also, carefully select the electrical, and HVAC contractors. The other trades will have little adjustment to make, so almost any competent crew will do.

The wall crew is most crucial. A crew with one experienced rough carpenter and one experienced concrete person works well. You can add to that an unskilled laborer to get a complete crew of three. If one person has both carpentry and concrete skills (as do many, who have both formed foundations and framed houses), add a semiskilled carpenter and a laborer. Usually, slightly adjusting the personnel of an existing carpentry or concrete crew is the easiest way to get one of these mixes of crew members.

Just as important, the wall crew should be interested in and enthusiastic about the new system. If they are, they will make the effort to learn the things they do not know, and they will take care with their work. If you have a wall crew with the right mix of skills, they can do the entire structure by themselves: footer (if any), ICF basement or stem walls, ICF exterior above-grade walls, floor decks, roof framing, and interior walls.

The electrical crew will have to cut many chases through foam and connect their boxes to unusual things (foam, ties, concrete), sometimes with unusual fasteners (screws, concrete nails, glue). This is not difficult, but they need to be willing to learn if they are going to avoid butchering the formwork the first time they try this. Make sure they do not do their foam cutting with whatever tool is closest at hand. Chisels and dull utility knives do a bad job. If necessary, consider lending them a router or hot knife (as described in chapter 2).

Sizing HVAC correctly is tricky because ICF walls are so well insulated and low in air infiltration. Most contractors will oversize the equipment, which misses the opportunity to save some initial cost. If possible, find a contractor who has worked with superinsulated houses before.

Materials ordering, storage, and handling

Most ICF manufacturers deliver within two weeks of receipt of an order. Plan the timing of your order accordingly. To avoid damaging the foam, store the units out of the way of work and try not to handle them frequently before use. In most systems the formwork is very light. One person can carry and move materials. Just before beginning formwork construction, you will need the ICF units inside the house perimeter. If you can safely sotre them there initially, that will save moving them later. Bear in mind that the grid panel systems are heavier. It takes two people or lifting equipment to carry the panels (as described in chapter 2).

If your units will be exposed for more than 30 days, protect them somehow from the sun, such as with an opaque tarp. Ultraviolet rays will eventually deteriorate the foam. This precaution is not necessary with the grid panel systems, however, since their special foam-and-cement material is nearly impervious to sunlight.

Site

The same soil preparation you use for the foundation of a frame house is usually adequate for an ICF house. However, the greater weight of concrete walls requires caution. If there is any concern that the soil might be soft or unstable, consult a local soils engineer or other qualified expert. It might be worth taking extra measures to settle or compact it.

If you plan to pour an ICF basement or stem walls with a chute, you will need to position the concrete truck at points around the perimeter about every 20 feet. Rough grade the lot accordingly.

Foundation

You can build ICF walls off of a conventional poured footing, a conventional poured or block basement, conventional poured or block stem walls, or you start off of a conventional slab. Regardless of the type of foundation, stress to the crew that it is important to maintain level. It is possible to correct for a moderately uneven foundation by adjusting the formwork above, but it is easier to get the footing or foundation right in the first place. When building off of a footer or slab, always form it; trenched pours are difficult to level accurately.

Whatever concrete is below your ICF walls, you will have short rebar (also called dowels) sticking up. Those bars will have to be placed in the concrete to line up with the ICF cavities above. That is not difficult if you are using any flat system (flat panel, flat plank, flat block). Your tolerance on the positioning of the bar is the same as for any conventionally poured concrete wall. But in grid and post-and-beam systems, the rebar is intended to go up the center of one of the cylindrical vertical cavities. Being off center by more than an inch or two can weaken the wall. More than 3 inches usually puts the bar out of the concrete altogether.

For accurate dowel placement in grid and post-and-beam systems, plan ahead. Find and follow the manufacturer's rules for rebar placement. If there are none, you can read chapter 5 to think through the position of each vertical cavity. Before pouring concrete for the foundation below, mark the exact position for each dowel on the formwork. Leave the dowels to one side. After pouring, wait a few minutes for the concrete to stiffen slightly, then push each dowel into place. There are ways to correct them later if some bars are out of position. But again, being accurate the first time is less work.

5

Setting forms

Because the design of ICF units and the methods of connecting them differ from system to system, the details of the setting procedure also vary. This chapter first presents step-by-step instructions for any of the block systems (flat block, grid block with fastening surfaces, grid block without fastening surfaces, or post-and-beam block). Even if you are using another type of system, these instructions introduce most of the basic procedures you will use. Later in the chapter are the changes you need to make to the block instructions in order to install the other systems.

Remember that many ICF manufacturers provide instructions for setting their forms. These take precedence over the instructions listed here. The directions here are only for planning purposes and to provide guidance when no manufacturer's instructions are available.

Setting block systems

Bear in mind one requirement of all block systems: The units must be set so that the vertical cavities align precisely all the way up the wall. If they are not, the vertical cylinders of concrete cast inside (sometimes called "posts") will have narrow points or even complete breaks. This is unacceptable because it will weaken the final wall. That is less of a consideration for flat systems since their cavities form one continuous slab of constant depth. However, the flat block systems all have vertical ties that act as fastening surfaces. These must align from block to block for ease of applying interior and exterior finishes. So their "vertical cavities" (the spaces between the ties) must align as well.

The first story of ICF formwork almost always rests on some type of concrete foundation. The ICF walls will be either basement or stem walls resting on a footer, or first-floor walls resting on conventional basement walls, stem walls, or a slab. They will then continue through the other stories up to the roofline.

Preassembly

As noted in chapter 3, you assemble the blocks of some systems before setting them. In addition, it is best to assemble all door and window bucks for the first story before beginning the walls.

Block assembly varies with the system. Consult the manufacturer's documents for instructions. Have enough assembled in advance to do a day's setting without interruption, or have the least-skilled crew members assemble blocks while the others set them. Store the blocks for the first story inside the house perimeter. This saves time and motion during setting.

Bucks are built to the door and window rough-opening size. The two variations are diagrammed in Fig. 5-1. The exact design depends on the exterior siding and type of door or window to be used. If you are siding with stucco, you will probably want to use a "stucco buck." It is designed for in-wall mount windows that are recessed, and it leaves a layer of foam on the outside to be stuccoed. In contrast, the "flanged buck" is better for flanged windows that are trimmed outside with a window casing.

The stucco buck consists of four sides (one sill, two jambs, one lintel) with temporary diagonal braces inside to maintain square. The lumber for the jambs and lintel is as wide as the maximum thickness of the concrete (see Table 3-1). For windows, the sill consists of a pair of 2-x-2s with a gap between for pouring concrete. For the bottom of a door buck, any size lumber will do; it will be removed later. Depending on your local code, you might need to use pressure-treated lumber for the jambs and lintel, or at least wrap it with tar paper. It is a good precaution to make the sills out of PT, even if not required.

For very large openings, put vertical and horizontal 2-x-4s inside to act as a brace against the pressure of the concrete during pouring. Attach all braces with screws or duplex nails for easier removal later. Finally, put fasteners into the outside perimeter of the buck, centered between the front and back faces. These will be cast into the concrete to hold the buck securely. For flat and grid block systems, use a duplex nail every foot. If you are concerned that the final door or window will be subjected to heavy stress, use a J-bolt instead at every corner and every 3 feet in between. For post-and-beam systems, a long fastener (a spike or a piece of rebar drilled into the lumber) is necessary to guarantee reaching the nearest post or beam.

This stucco buck will mount between the front and back faces of the foam formwork as shown in Fig. 5-2. An in-wall mount window will slide into the buck and attach to it with screws through its frame.

When using doors and windows with flanges, the "flanged buck" is necessary. The flanged buck begins just like the stucco buck, except that the jambs and lintel use lumber exactly as wide as the total width of the blocks (as listed in Table 3-1), and the sills are made of parallel 2-x-3s or 2-x-4s (whichever leaves a gap between them of 3–4 inches). On both the inside and outside face of this frame go flanges of ¼-inch plywood or almost any scrap wood, 6 inches wide,

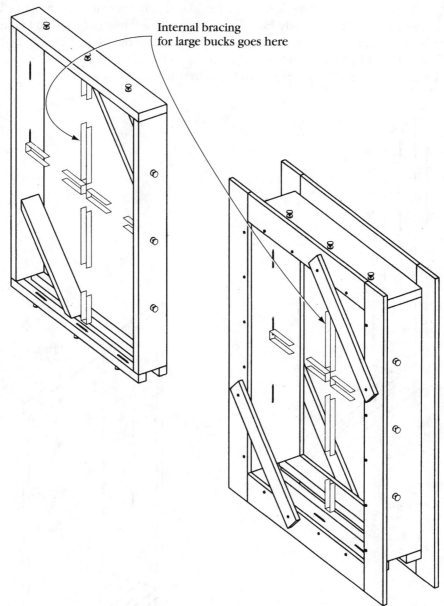

Internal bracing
for large bucks goes here

5-1 *A stucco buck (left) and flanged buck (right).*

and then some diagonal braces to maintain square. (On a door buck, put no flanges on the bottom or along the bottom 2 inches of the sides.) Nail or screw the flanges every 8 inches; use removable fasteners on them and the braces. Add fasteners to cast into the concrete, as with the stucco buck. Add vertical and horizontal braces inside for large openings, as with the stucco buck.

After removal of the flanges and bracing, the rough opening from a flanged buck (drawn in Fig. 5-2) is ready for a flanged window and trim to be nailed to the outside face. It can be used with stucco siding, but then it must either be trimmed conventionally, or else the exposed buck lumber and window flange must be covered with foam or special tapes so that the stucco will stick to them properly.

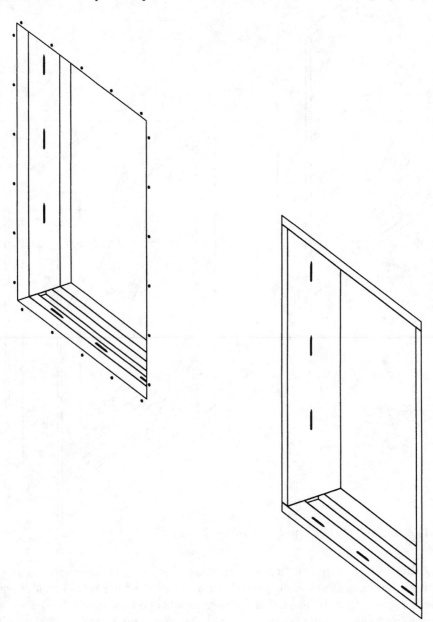

5-2 *A stucco buck (left) and flanged buck (right) as they appear mounted in the formwork.*

Foundation prep

Figure 5-3 depicts a foundation that is being prepared for setting formwork. Snap chalk lines on the foundation (footer, slab, or walls) on which the ICF walls will stand, one for the inside surface of the blocks and one for the outside. Then nail (with concrete or cut nails) 2-×-4s just inside the inner lines and outside the outer lines. These so-called "guides" will hold the bottoms of the blocks in position during construction. If the blocks will line up with or overhang the foundation below, you will have to mount the guides onto the face of the foundation instead of the top. Put them around the entire perimeter. Where a door buck goes, erect the buck and plumb it with diagonal 2× braces (also called "kickers") attached at one end to the buck and at the other to the floor or a stake in the ground. Finally, on the guides or the foundation, mark the points along the wall where each window is to go.

Making corners

Begin the first course by setting the corner blocks. Figure 5-4 shows this job partly completed. If you are using a post-and-beam system, bear in mind before you start that for each course of blocks that is to be filled horizontally with concrete (a "beam"), you will need to use the special lintel block, which allows con-

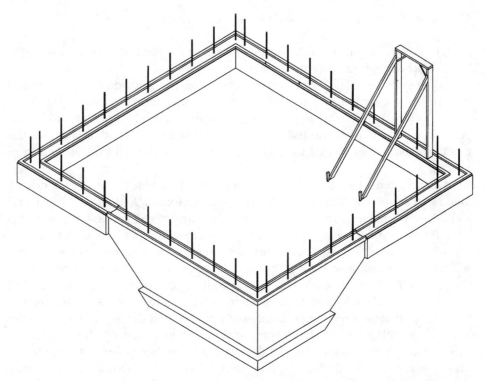

5-3 *A foundation (part footer and part stem wall) that has been prepared.*

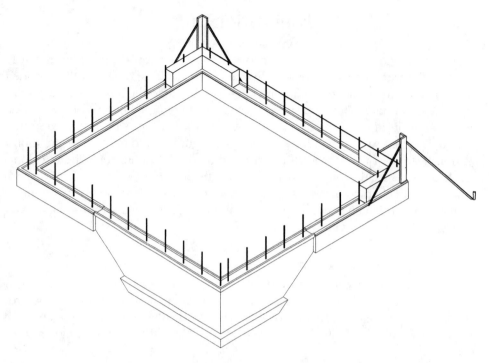

5-4 *Some of the corners assembled on the foundation.*

crete to flow horizontally. So if your design calls for the first course to be lintel
block, find enough now for the first course.

Some systems include "premolded" corner blocks, and some use a standard
or "end" block butted into the corner like a conventional concrete block, as
shown in Fig. 5-5. For the second ("butted") type, you might need to add foam
end pieces or cut away the teeth from the end blocks. For either type, set the
block at each corner so that its long end sticks out to the right (as viewed from
inside).

Other systems include "precut" corners (standard blocks cut at a 45-degree
angle so that you can abut and glue two pieces to form a 90-degree corner), and
still other systems require that you cut standard units to assemble your own corners. Precut corners generally come in two versions: long in both directions and
short in both directions. If your system requires cutting your own corners, there
will usually be instructions from the manufacturer. If not, you can generally use
the procedure diagrammed in Fig. 5-6 to produce your own "longs" and shorts."
Choose a vertical cavity just off the center cavity. Cut two blocks down the center of that cavity at a 45-degree angle, but cut one with the angle slanted to the
left and the other with it slanted to the right. Then match up the two longer
pieces and the two shorter ones and glue them together to form 90-degree angles. You can use a little duct tape to hold the corners together while the glue
dries, or use more duct tape and no glue at all. Whether you are using precut

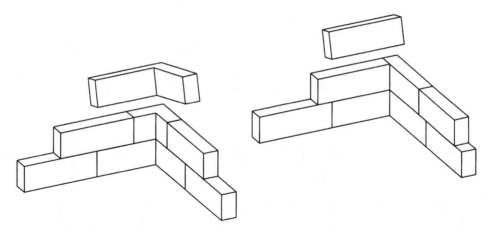

5-5 *The setting pattern of premolded corner blocks (left) and corners constructed of "end" blocks (right).*

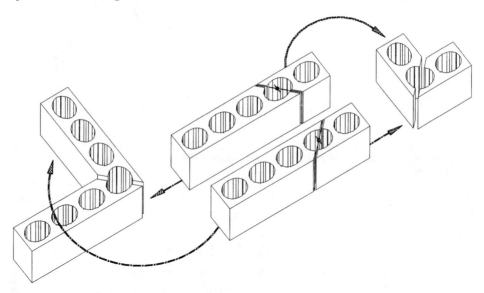

5-6 *Construction of long and short corners from two standard blocks.*

corners or making your own, set the blocks that are long in both directions at the corners of the foundation.

Now construct and install a "corner brace" outside each corner block (see Fig. 5-7). A corner brace consists of a 2-x-6 and a 2-x-4 nailed along their edges to form a sort of L-channel. Stand its end on the guides, plumb it with 2 kickers that run to the guides and another to the ground, and toenail it to the guides. This will keep the corner plumb during setting and pouring.

If your corners are precut or otherwise assembled from more than one piece, you will need braces inside each corner as well. Make another brace for

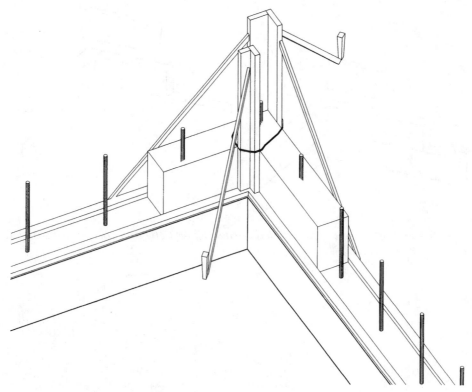

5-7 *An outside corner brace (rear), wired to an inside corner brace (front). The wire and inside brace are necessary with only some types of corners.*

each corner, set it inside the blocks, and tie it and the outside brace together with wire just above the blocks, as in Fig. 5-7.

Setting the first course

Pick any wall to begin with. To set a level for the entire course, attach the ends of a string line to the two end corners at the height of the top of the corner blocks, as in Fig. 5-8. Set a standard block next to the corner block on the left, and measure its deviation from the string. Shim it or shave it as necessary to get it to level. Take it out again and run a thin bead of glue along the left edges of the front and back faces of foam (see Fig. 5-9). Then replace it and butt it tightly to the corner block until the glue grips. Continue setting blocks in this fashion, working left to right.

Note that some blocks are designed to stick to one another by the friction of their interlocking edges without glue. If you are using such "friction-fit" blocks, you need not put glue on the blocks' edges to connect them. That applies to all higher courses of block as well as the first.

Each time you reach a dowel coming out of the foundation, you will thread the block over it. If you find that any of these dowels is far off the center of the

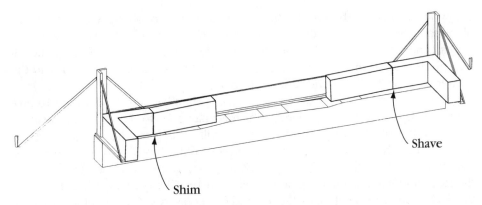

Shave

Shim

5-8 *Adjustments to level the first course of block.*

5-9 *Gluing the edges of a block.*

vertical cavity it is supposed to fit into, you will have to correct it. Remove the block and do not glue it yet. Cut off the dowel with any metal-cutting blade, as far down as possible. Drill a hole into the foundation in the proper location, sizing it ⅛" larger than the rebar diameter. It is generally safe to drill to the same depth as the other dowels, but check with an engineer or the ICF manufacturer to be sure. Drop a new piece of rebar into the hole, securing it with a rebar epoxy or other suitable adhesive. Then continue setting.

When you reach the end of the wall, you will probably have to cut the last block to fit precisely to the corner you set previously. Cut it ¼" short, glue its left edges (unless you are relying on friction to hold the blocks together), slide it into place snug against the block to the left, and fill all gaps between it and the corner block to the right with adhesive foam.

When you reach a door buck, you will also cut a block to fit. Figure 5-10 provides details for a window opening, which is analogous to a door. If it is a stucco buck, cut the block so that the foam faces will exactly overlap the jamb of the buck. You might also have to shave the inside of the block's cavity so that it fits over the jamb. Glue the block along its left edges (if necessary), set it in place over the jamb, and foam nail it to the jamb every 8 inches. If it is a flanged buck, cut the block ¼" short of the jamb, glue the left end (if necessary), slide the block between the flanges into place, and fill the gaps between its right edges and the jamb.

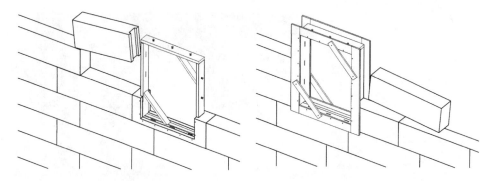

5-10 *Setting blocks to a stucco buck (left) and flanged buck (right).*

To resume setting on the opposite side of a door buck, the next block must be cut so that its right end will fall precisely where it would have if the buck had never been there. Otherwise the vertical block cavities could go out of alignment above. To do this you can simply measure from the last joint between blocks before the buck. Figure 5-11 depicts the measuring points. Since the block joints fall at regular intervals, you can find the distance from the buck itself to the end of the next block to the right. This is the length to cut the block. Glue its left edges, place them snug to the buck, and resume setting.

Set the first course of the next wall to the right in the same fashion. Then continue around the perimeter left to right (clockwise) until all walls have a first course. If you have enough crew available, you can set block on more than one wall at a time. Start each setter on a separate wall, working from left to right.

Finishing the first course

A few steps are necessary to complete the first course. Figure 5-12 shows them in progress. Wherever there is a gap between blocks and concrete (such as where the block is shimmed up), fill it with adhesive foam.

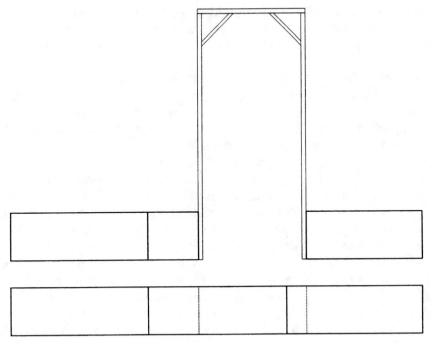

5-11 *Cuts to make in blocks (below) to set around a door buck (above).*

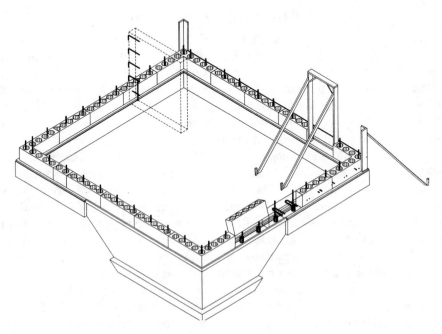

5-12 *First course of block that has been finished.*

Now you will place a collar around each dowel to receive vertical rebar when the wall is fully set. Cut a 2-inch diameter PVC pipe into 3-inch lengths. Thread one of these collars over each dowel so that it rests on the foundation below.

Put any embedded fasteners in next. These will be primarily for interior walls and particular sidings. Interior walls can be attached with other methods later, but if you know where they will go and you prefer a concrete connection, you can push duplex nails (for lighter walls) or J-bolts (for load-bearing walls) out through the block to catch the end studs later.

Sidings can be attached to fastening surfaces later, so you only need embedded fasteners for siding if your system has no fastening surfaces. Procedures for embedding fasteners for sidings are as follows.

If you will be siding with PC stucco, push a U-shaped wire through the foam about every 16 inches horizontally and vertically. The U-wires will hold the stucco's wire mesh to the wall. If you will side with a brick veneer, install brick ties according to the same procedure recommended for U-wires. However, check the proper spacing of the brick ties first. If you will use a nailed siding (clapboard, shingles) or screwed siding (vinyl, aluminum), you might want to make provisions for furring strips. (If you do not, you can attach them later with some extra effort.) To do this, push through duplex nails where you want to attach the strips. One nail every 2 feet vertically should hold the strip.

Finally, if the design calls for rebar in the first course, set it. Rest it on the ties inside the block. Remember to lap the ends of adjacent bars.

Setting higher courses

The courses after the first are set in the same way, with minor modifications. Figure 5-13 shows setting up a few courses. If using a post-and-beam system, before each new course check whether the course is to be filled with concrete. If so, build it out of lintel block. As you set each lintel block, also remember to puch out or fill holes in it as appropriate. You must have a hole in the bottom of the lintel block over each vertical cavity that is to be filled with concrete below. The concrete will flow down through the hole, forming a continuous post.

For the second course, begin again at the corners. The vertical joints of the blocks need to be staggered from course to course (just as on a conventional block wall) for strength. If the first-course corners were long to the right, set second-course corners that are short to the right; if they were long in both directions, now set corners short in both directions. Remember that at all times the vertical cavities of the blocks must line up, so construct the corners accordingly. Glue the corner blocks continuously along their bottom edges (if necessary) before pushing them into their final position. If you have corner braces on the inside, wire them to the outside braces just above the new corner blocks again.

Work left to right around the perimeter as on the first course. Glue each block (if necessary) on its left edges and its bottom edges so that it adheres to

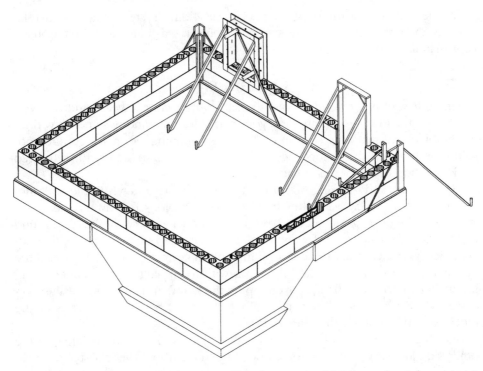

5-13 *Progress after a second course of block: staggered joints and notches cut out for bucks.*

both the block to the left and the blocks below. When cutting a block to fill the end of a wall (abutting the right-hand corner), be certain that the block is cut so that its cavities line up precisely with those of the blocks below. Glue it on the bottom and on the left end (if necessary) before setting into final position. Fill the gap between the right end and corner block with adhesive foam.

If at any time you notice that the vertical cavities of a block are out of alignment with those of the blocks below, stop setting. Go to the beginning of the wall (far left corner) and move to the right until you find the first point at which the cavities go out of alignment. Pull out the top course of block to the right of this point and replace them so that the cavities do align correctly.

When the second course is set, finish it as you did the first. Continue setting higher courses. If you are using a post-and-beam system, remember before each course to determine whether you need lintel block. Remember also that the courses at the level of the floor decks must be lintel block to anchor the deck into. For all systems, alternate the type of corners used on each course to stagger the vertical joints. Wire the inside corner braces (if any) to the outside braces between each course, or at least every 16 inches vertically. Finish each course as described previously. Beyond a height of 4–5 feet, you will need something to stand on to continue setting. Set up fixed scaffolding or whatever else you prefer. Some ICF manufacturers offer special scaffolding systems that fasten relatively easily to their forms. These systems often serve as bracing as well. Contact your manufacturer for details.

Openings

When your wall first extends above the sill level of your lowest windows, you need to set the bucks for them. Find your marks for the windows and draw lines on the blocks to mark the bottom of the rough opening. For a stucco buck position the lines for the inside edges of the buck's sill and jambs; for a flanged buck, position them for the outside edges.

Cut the marked opening out of the block to create a notch in the wall into which to set the buck. Save the cutout sections of block for further use. For a stucco buck, you might need to shave inside the cavities of the blocks around the opening to make room for the buck to fit. Lift the buck into place, then attach kickers as necessary to hold it plumb and in correct position. Then nail the inside and outside faces of the block to it with foam nails. For a flanged buck, lift the buck into place (flanges overlapping the foam) and attach kickers as necessary to plumb. Then fill any gaps along the sides between the ends of the blocks and the jambs with adhesive foam.

On higher courses, set blocks up to the window bucks and resume on the other side just as you would for a door buck (see above). For a stucco buck, foam nail each block to the buck as you set it. Make sure you have one nail near the top of each block and another near the bottom, as well as one at least every 8 inches in between.

When setting the first course of block that extends above the lintel of an opening, you will need to cut blocks to fit over the buck. Figure 5-14 shows how this is done. If you have saved the block cutouts from the bottom of a window, some or all of them can often be used above, reducing cutting and waste. Regardless, cut blocks to keep the level of the course constant. (Do not simply stack whole blocks on top of the buck). Also keep the vertical cavities aligned. Foam nail the blocks to stucco bucks, and attach the blocks' edges to the jambs of flanged bucks with adhesive foam.

When the course that extends above a lintel is done, be sure to put the required rebar above each opening.

Setting vertical rebar

When you have set the top course for your first ICF story, set the vertical rebar. Figure 5-15 shows a vertical bar as it would be set in most systems. Slide each bar down the appropriate vertical cavity. Guide its bottom end into the collar around the dowel below. When the bottom is secure you will need to secure the top so that the bar does not fall against the front or back face of foam during the pour, or against the ties. Cut a long piece of wire, wrap one end around the tie to the left of the bar, the middle around the bar, and the other end to the tie to

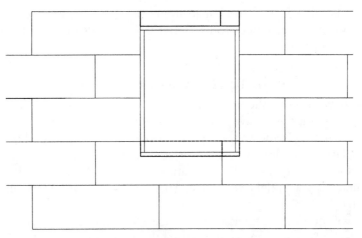

5-14 *Pattern of cutting out block at the bottom of an opening and reusing the scrap above.*

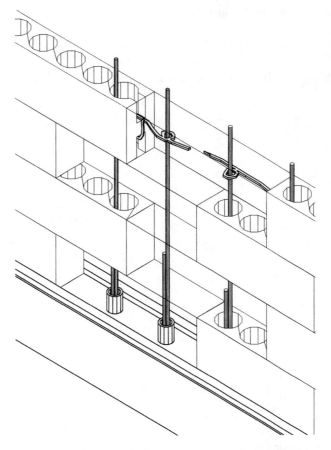

5-15 *Securing vertical rebar above and below.*

the right. This will suspend the bar in the middle of the cavity. If the ICF will continue up further, remember to use rebar that extends beyond the top of the wall to lap the bars above.

Finishing the top of the story

When you have reached the top of your first ICF story, how you finish the top depends on whether you have just set your top story or you will continue up another one. If this is the top story, the top of the formwork should extend above the planned roofline. Use a transit or similar tool to find the exact roofline and mark it all the way around the perimeter of the building. Cut off the excess block, maintaining a straight, level line. This is all that is necessary until you do final bracing.

If you will build up further, you will need to make provisions for attaching the floor deck. There are two common methods of connecting a conventional floor to ICF walls, called the "pocket" and "ledger" methods. Figures 5-16 and 5-17 contain details. Either method will work in almost any situation, but they have different advantages that might make one preferable or necessary in particular cases.

Under the pocket method, you place wooden blocks into the formwork before the pour so that they form recesses (called "pockets") into which you can lower floor joists. This is simple and uses the minimum of material. It brings

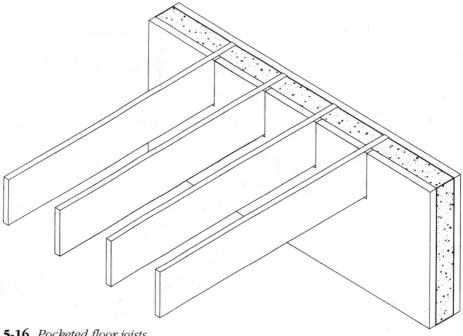

5-16 *Pocketed floor joists.*

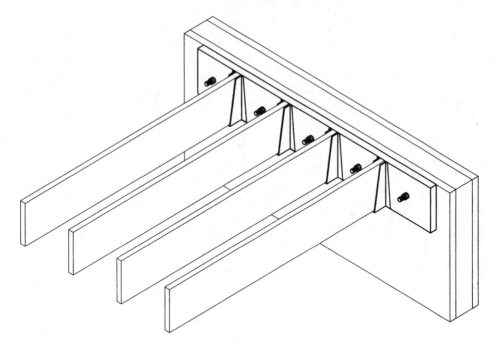

5-17 *A ledger and attached floor joists.*

each joist in direct contact with concrete, so your local building department officials might require the lumber to have a water-resistant coating over its ends, or perhaps even be all pressure-treated. This method also limits the positions of the joists to the pockets. There is little "wiggle room" and no easy opportunity to make modifications to the deck design in the field.

Under the ledger method, you attach a wooden ledger board along the walls that will receive joist ends. The joists attach to the ledger with joist hangers. Only the ledger is in contact with concrete, so only the ledger needs to be protected from moisture. You can use PT lumber for the ledgers or use KD with plastic sheet tacked to the back face. The joists can also be attached anywhere. However, more material is used (ledgers and joist hangers), and construction takes more time.

Figure 5-18 shows the pocket method in progress. If you use this method, build the ICF wall up close to the design level of the top of the joists, or even a little higher. If the wall is not well above the level of the bottom of the joists, the pockets will not be deep enough to hold the joists steady during deck construction. Snap a chalk line precisely at the design level of the bottom of the joists. Cut a set of wooden blocks about 8 inches long and high enough to extend from the chalk line to the top of the form wall. Cut slots in the form wall to hold the wooden blocks at precisely the positions the joists are to go.

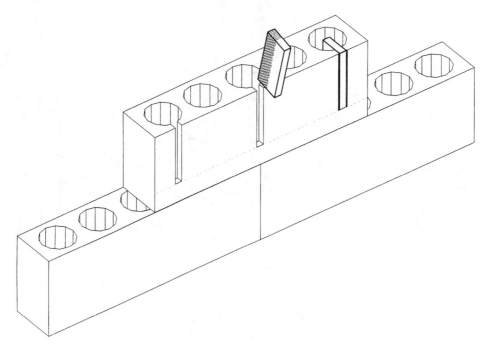

5-18 *Inserting wooden blocks for forming pockets.*

Use rough-sawn or true-dimension lumber for the blocks so that they will be slightly thicker than the joist ends. This will allow the joists to slide into the pockets easily. Coat the ends of the blocks that will protrude into the form cavities with oil. Form oil works well if you have it. This will make it easier to knock the blocks out later. In many areas, code calls for 3 inches of joist bearing on the end walls, so the blocks would have to protrude at least 3 inches in to the formwork cavities. Required bearing is different in some areas, however. Glue the blocks firmly to the slots in the formwork so they do not move or allow concrete through during the pour.

Figure 5-19 depicts the ledger method in progress. If you use this method, build the wall up to or past the design elevation of the top of the floor deck. Place a chalk line at the design elevations of the top and bottom of the joists. Between these is where the ledger will sit. In every vertical cavity, cut out a rectangle of foam from the formwork that ends about 1 inch short of the chalk lines (at top and bottom) and about 1 inch short of the sides of the cavities. These will allow the concrete to flow up to the ledger.

Place J-bolts in the ledger boards so that they will go through the cutouts and into the approximate center of the form cavities. If your system has fastening surfaces, screw the ledger boards in place. If it does not, run threaded rod through the ledger boards, through the foam ties, and out the opposite side of the formwork. Make sure you drill your holes in the boards so they line up with

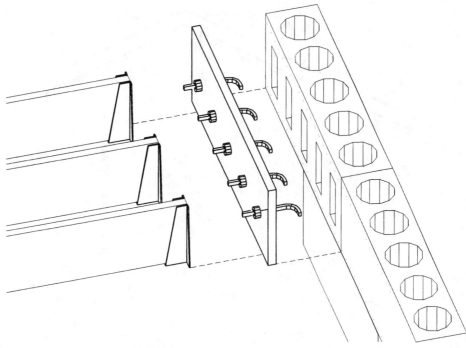

5-19 *Assembly of a ledger floor.*

the foam ties; otherwise the rods will lock into the concrete and you will be unable to remove them after pouring. Screw large washers and nuts on the ends of the threaded rod to hold the ledger to the formwork.

Final bracing

Before placing concrete you need to add some additional bracing. Figure 5-20 depicts this bracing in progress. The top edge of the wall needs bracing to keep it from spreading out during the pour and to provide a nailing surface for additional kickers. Figure 5-21 shows this. If your system has fastening surfaces, screw 2-x-4s on the surface of the formwork along the top edge, both inside and out. If this would interfere with a ledger that has been installed on the inside, skip the inside of that section of the wall. The ledger boards will serve the same purpose. If you have no fastening surfaces, construct "ladders." Use parallel 2-x-4s connected by a piece of strapping every 2 feet. If a ledger interferes with a ladder, nail the strapping to the ledger, using it as the inside rail of the ladder instead of 2-x-4.

At the right-hand end of each wall will be all the odd cuts of block made to achieve correct wall length. Place against each side of this section of wall a piece of plywood that runs the height of the wall and is wide enough to cover the seams of the cut blocks. Screw them to the fastening surfaces if your system has

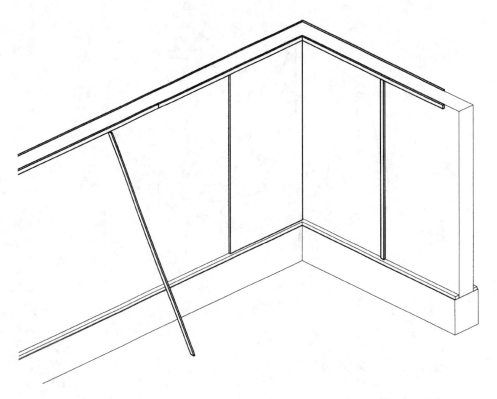

5-20 *Wall kicker (left), corner plywood (center), and vertical 2-×-4 brace (right) as final wall bracing.*

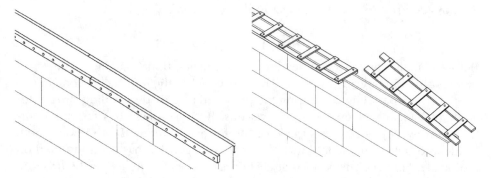

5-21 *Bracing the top of the formwork with lumber fastened to the block's fastening surfaces (left) and with ladders (right).*

any. If it does not, run a couple of vertical 2-×-4s between the guide below and the ledger or ladder above to hold the plywood flat against the wall surface.

Some manufacturers recommend intermediate bracing for long stretches of wall without openings. The intermediate braces are usually 2-×-4s set vertically against the wall between the guides and the top edge bracing, as described previously. You can add this if your manufacturer recommends it, or anywhere you believe you have a weak spot.

Finally, double-check the wall for plumb. You can adjust it as necessary by moving the kickers attached to the bucks. Where you have a long stretch of wall without any bucks, run an occasional kicker to the bracing along the top edge to plumb the wall.

Second and third stories

In a multistory house of ICF walls, you will fill the first story with concrete (see chapter 6) and then set the formwork for the second on top. Thus you will be setting the upper stories on top of other formwork instead of plain concrete. The procedure is identical to the first story except for a few minor modifications in the early steps, shown in Fig. 5-22.

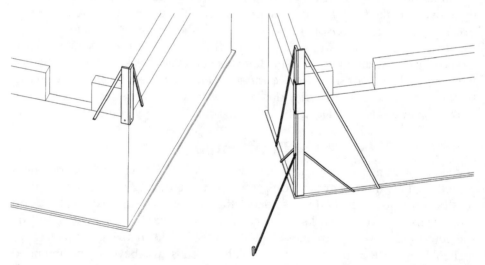

5-22 *Upper-story corner braces fastened to the block's corner braces (left) and set on the lower-story corner braces (right).*

Some times plans call for a different wall thickness on different stories. Usually this change occurs between the basement and first floor. If you wish to do this, consult the manufacturer. Many ICF manufacturers provide details for a transition between two different widths of their units.

If your system has fastening surfaces, screw exterior corner braces directly to the formwork. Position them so the bottom laps the lower story by a couple of feet for stability. Plumb them and attach diagonal kickers to the wall's fastening surfaces to secure their position. If you have no fastening surfaces, leave the exterior corner braces below in place and set the new ones on top. Plumb with kickers to the ground.

In most cases the top of the formwork from the story below will extend a few inches above the floor deck on which you are now working. When you set door bucks, you will therefore need to cut a section out of the block to make room for them.

Set the first course of block directly on the top course from the story below; there will be no guides. Start at the corners (as usual). Stagger the types of corners you use with the corners immediately below, and align cavities with the cavities of the story below.

When setting the corners, place them in position without gluing first. Run a string line between adjacent corners. Check the strings for level. Also measure down from the string to the blocks below at a few points between the corners to check for level along the wall. If everything appears to be reasonably close to the same elevation, simply remove the strings and forget about further leveling. Any unevenness will be taken out when you cut off the roofline. But if things are so far out of whack that they threaten to weaken the interlocking of blocks up above or make setting time-consuming, then shim up or shave the corners as necessary to achieve consistent level, and leave the strings up for later use.

If the course below was nearly level, set the rest of the course just as you would any course after the very first. If it was not level, check each block against the string and shave or shim as necessary to get to the correct height. Shaving can be tricky because of the interlocking teeth or grooves of the blocks. You must shave both the bottom edge of the new block and the tops of the blocks below.

Roof ends

If you are building gable or gambrel ends out of your ICF, you will handle them something like another upper story. When you are cutting off the form wall at roofline level, leave the end walls uncut. After the pour, resume setting block on these walls. No corner blocks are necessary, but all vertical joints must still be staggered and all vertical cavities aligned. Set the blocks in a ziggurat pattern, as in Fig. 5-23, so that the wall at all points extends just beyond the intended roofline. Snap chalk lines along the formwork at the height of the roofline, allowing for the thickness of the top plate (see below).

If you are using a post-and-beam system, you will also need to allow for the height of an extra course of lintel block, as shown in Fig. 5-24. The gable ends must have posts and beams running in the end walls, plus beams running on the diagonal along the roofline. So cut the formwork shorter by the added height of a course of lintel block set on the slope. Finish off the wall with the lintel block, and of course set rebar in it before pouring. Normally the rebar pattern used in other walls is sufficient in the end wall, and the same amount of rebar used in an ordinary beam is sufficient for the lintel course on top, but check with the manufacturer's literature, the manufacturer, or an engineer to be sure.

Before placing concrete, cap the wall with top plates, as shown in Fig. 5-25. You can use 2x plates as wide as the maximum width of the concrete (see Table 3-1), shave the inside of the cavities to make room, and set the plates

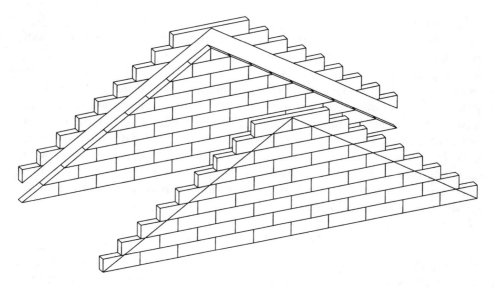

5-23 *Stacking block for a gable end (bottom) and cutting it to the roofline (top).*

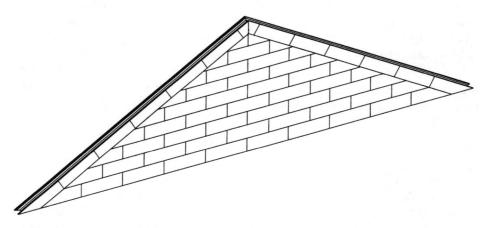

5-24 *Gable end formwork with a course of lintel block along the top.*

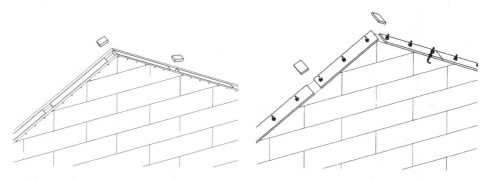

5-25 *Inserting top plates into a gable end (left) and gluing them on top (right).*

inside the block and foam nail it in place, much like a stucco buck. This is preferable if you intend to put stucco siding right up to the underside of the roof. Or you can use plates as wide as the total width of the blocks (Table 3-1) and glue them on top. This is preferable for nonstucco sidings, or if you intend to run trim under the roofline. Regardless, install the J-bolts into the plates before setting them, holding the bolts to the proper depth by screwing a nut down on either side of the plate. Also leave a gap of 3½ inches (the true width of a 2-x-4) between the ends of adjacent plates. Set the length of these gaps by physically inserting a short piece of 2-x-4 between the plates while installing them. The gaps will give you access during the pour; you will fill each gap with a wooden block at the right time. Run kickers to the openings (if any) and top plates to plumb.

Irregular corners

The AAB system has optional "hinged" blocks that can be set to form corners of virtually any angle, and other manufacturers might be offering something similar in the near future. Otherwise you make non-90-degree corners in the same way recommended to make 90-degree corner block from standard units (discussed previously in this chapter under "Setting corners"). You simply cut the blocks at half of the planned wall angle. (Cut at 30 degrees for a 60-degree corner, and so on.) As for 90-degree corners, make both "longs" and "shorts" so you can stagger vertical joints. The guides for the first course should follow the irregular angle. Set the corner block along with the other corners. The corner braces will also need to be built with the same irregular angle to them.

Curved walls

It is relatively easy to make accurate radius walls by building your own radius blocks. Cut a vertical slot out of the inside face of each cavity of the block, as in Fig. 5-26. Glue the inside edges of the slots, bend the block to close up the slots again, and hold in place with duct tape or some form of homemade jig until dry.

The wider the slots cut out, the tighter the radius. Some manufacturers publish tables showing how wide a slot to cut for different radii. If you have no table, you will need to experiment. In any case, cut all slots the same width. For blocks with a half cavity at each end, cut a "half slot" on each end. (That is, trim the end by one-half the width of the other slots.)

Since the vertical joints of the blocks do not line up, to get the curve to start at a consistent vertical line in the wall you will have to make a few blocks curved only on the right-hand end and a few curved only on the left. Play with this on paper or at the job site to figure it out. Note that the guides on the foundation will have to curve with the wall, too. Simply make the guides out of short pieces of 2-x-4 set at angles, or out of 2-x-10 or 2-x-12 cut on a curve with a jig saw, to follow the planned wall line.

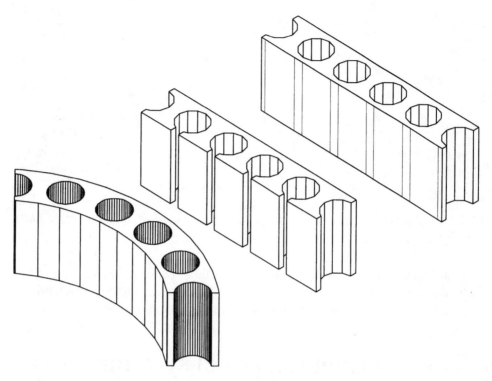

5-26 *Cut lines (upper right), cuts (middle), and final shape (lower left) of a block cut to curve.*

Irregular openings

To make window and door openings with curves or odd angles, do not build a buck in advance and stack around it. Instead, stack the wall solid with block. Then mark the opening on the wall and cut it out. Build a buck to fit and slide it into place. If you need flanges, you can put them on one side before sliding the buck in, and attach those for the other side after.

If the opening has curves, you can build the buck with a series of straight pieces that approximate the curve. Or you can curve a piece of thin lauan underlayment or other sheet material.

Shelves

You might plan to cantilever some of your floor joists past the exterior walls (for example, to construct a projection such as a garrison or bay). To do this, build the story below as usual. When marking the elevations for the bottom of the floor deck, also mark the sides of the planned cantilever and cut a notch in the foam for it, as in Fig. 5-27. Glue wooden blocks to either side of the notch to act as a stop when the concrete is placed.

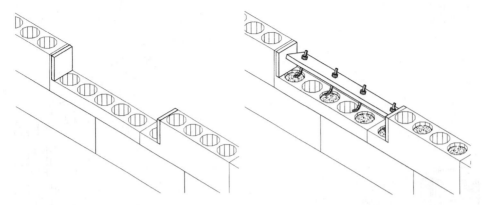

5-27 *Shelf prepared in the formwork (left) and detail of attachment of plate (right).*

During the pour, treat this shelf like a roofline. Even out the concrete flush with the top edge of the formwork and insert J-bolts. After the concrete hardens, bolt down a top plate. Now the shelf is ready for setting the joists.

Setting flat-panel systems

Setting flat panels differs in some details from setting blocks. Each panel covers more wall area than a block, and joints are connected with wire or special plastic fittings rather than glue or interlocking teeth. There is no need to stagger joints, although the plastic ties should align to make attachment of finishes (nailed or screwed sidings, interior wallboard) easier. In some of these details the two flat-panel systems (R-FORMS and Styroform) are different from each other. R-FORMS has available bracing systems that replace guides and site-constructed lumber bracing. Styroform uses the same type of bracing as block systems.

Preassembly and foundation

R-FORMS is assembled in the field from sheet foam and special plastic ties. Follow the manufacturer's instructions. Styroform is preassembled. With R-FORMS you buy your own foam, and with Styroform you can ask for different foams from the manufacturer. Use either XPS or EPS of at least a 2-pound density. A 1.5-pound foam is potentially too weak.

R-FORMS uses no guides. Instead, place two short pieces of U-shaped, 2¼6" metal channel, one just inside each chalk line, every 4 to 6 feet (see Fig. 5-28). Concrete nail it to the foundation. It will hold the panels in place until bracing. For Styroform, use guides just as with block systems.

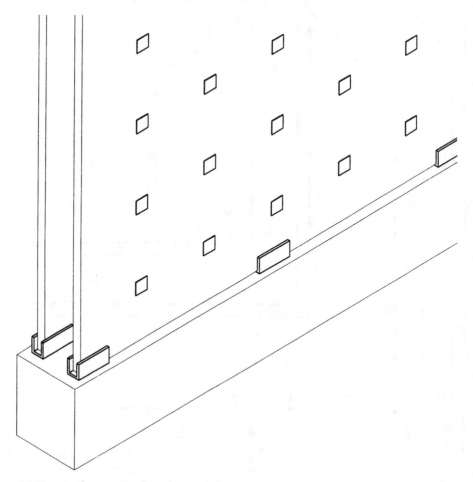

5-28 *Panel set in steel U-channel.*

Corners

Make corners by cutting the inner sheets of foam of two panels, as in Fig. 5-29. The inner sheet of each panel is cut short by the maximum thickness of the concrete (see Table 3-1), plus the thickness of one sheet of foam.

Set a corner by butting two panels together as in the figure. Then connect them with special plastic fasteners, as specified by the manufacturer. R-FORMS has available special preassembled bracing racks made of steel studding. Place these against the corner panels as you set them, as specified by the manufacturer. Place a standard outside corner brace behind each Styroform corner. There is no need to set all corners first. After the first corner is set, set the first course of panels around the perimeter in a clockwise direction, cutting and setting each corner as you encounter it.

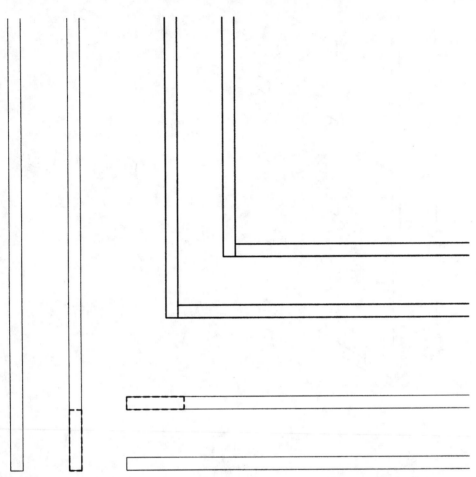

5-29 *Cut lines for panels at a corner (left and bottom) and assembled corner (upper right).*

Setting the first course

Shim or shave the panels as necessary to achieve level off of the foundation. Fill any gaps between panel and foundation with adhesive foam.

Except in odd situations, set R-FORMS on end (so each panel extends up 8 feet). Continue to set the R-FORMS bracing with each panel. Set Styroform panels on their sides. Attach adjacent panels with the plastic connectors provided by the manufacturer. With Styroform, also strengthen and tighten the connection with wired looped around the tie ends of adjacent panels, as in Fig. 5-30. Work clockwise around the perimeter. If you have two crews, have the second one work counterclockwise from the same original corner.

When reaching an opening, cut the panel end flat to the side of the opening, rather than cutting a hole or notch for the buck. See Fig. 5-31 for a detail. Cut

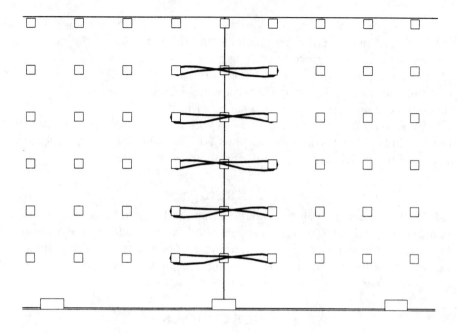

5-30 *Wire over the seam between two adjacent panels.*

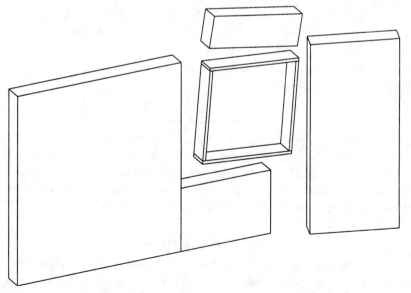

5-31 *Setting panels around an opening: the left and bottom panels first, followed by the buck, top panel, and right panel.*

and set a separate piece to go below the sill (for a window opening). Then slide the buck into place, start a new panel to the opposite side of the opening, and cut and set a separate piece to go above the lintel. Connect all adjacent panels.

Since the panels are large, horizontal rebar might need to go in the middle of a unit, not just along the top. After a panel is in place, but before you set the next adjacent one, slide the horizontal rebar in the side.

Embedding fasteners inside the formwork also requires more reach with the larger panels. Often it goes more easily if you push them through the foam while you have the panel on the ground, and then set the panel.

Bracing

R-FORMS' bracing is set along with the panels and should be complete when the panel setting is done. Plumb it as specified by the manufacturer. Styroform bracing is about the same as for block systems: 2-x-4s or a ladder along the top and kickers to the bucks and the top bracing. No plywood is necessary along the corners or over seams in either system.

Later courses

The 8-foot-tall R-FORMS panel often reaches the level of the floor deck, so no second course is necessary before the pour. In the case of a high ceiling, consider using longer sheets of foam or placing the forms sideways and setting three courses to make a 12-foot wall. You can also make a second course out of panels cut to some intermediate height, such as 2 feet, to stack on the 8-foot first course.

With either system, make each higher-level corner exactly the same as the first. Set corner and other panels directly over like panels below. This will align all vertical joints. As noted previously, it is unnecessary to stagger joints for strength reasons. Aligning them keeps all cuts the same from course to course and automatically aligns the ties.

Roof ends

Especially for the largest panels, it is easier to precut the units that will be used for gable or gambrel ends. This reduces the size of the units that must be lifted into place, and it makes it easier to incorporate the cutoffs elsewhere in the wall (rather than wasting them). Precut the end wall panels slightly oversized, then pull the chalk line and do final cutting in place, as with blocks.

Irregular corners

Form non-90-degree corners by cutting the inside foam sheets of the corner panels slightly short. The more you cut off the sheet, the sharper the resulting wall angle. The manufacturers can provide details for connecting panels at a non-90-degree angle.

Curved walls

You can form curved walls in the same fashion as with blocks: Cut vertical slots in the inner sheet of a panel, glue the edges of the slots and bend the panel to join them. However, the greater size of the units makes it more difficult to bend them. You might prefer to cut completely through the panels vertically to form a series of beveled flat panels that approximate a curve (as in Fig. 5-32) and glue these together.

Setting grid-panel systems

The grid panels are made of a heavier and stiffer material than the other ICF units. For this reason, there is more use of lifting equipment and less need for bracing. The following are the details of setting that differ from the instructions for block. The two systems (ENER-GRID and RASTRA) are similar to one another; no separate instructions are given.

General setting procedure

You will generally set the panels horizontally. The only exception is narrow vertical wall sections, such as the space between two adjacent doors. In these situations it might save cutting to stand a panel on its end. In all cases, however, you must cut and place units so as to keep both vertical and horizontal cavities aligned from panel to panel.

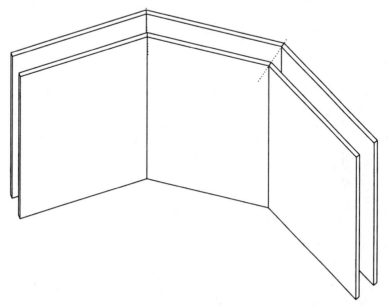

5-32 *Approximating a curve with beveled cuts along the edges of a panel.*

Corners and first course

It is unnecessary to put guides on the foundation. Simply snap chalk lines to mark the inside and outside of the ICF walls. To begin setting, cut through the centers of the end cavities of two panels at 45 degrees, and abut them to form a 90-degree corner (see Fig. 5-33). Set them temporarily in place and check for level. Shim or shave as necessary to achieve level. Glue them continuously along their ends and set them in place. Also glue continuously along their joints with the foundation below, both in front and in back. Then set adjacent panels to the right, shimming or shaving to level. And glue each one on the left end and along the bottom. At each corner, make a 45-degree cut in a panel to end the wall, and in another one to start the next wall. The smaller panels (15 inches

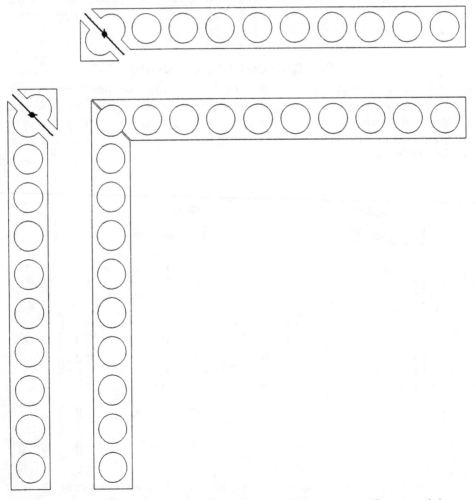

5-33 *Cuts in ends of grid panels (left and top) to form a corner (lower right).*

wide) can be set in place by two workers. The double-wide panels, and sometimes small panels that must be set high, require lifting equipment.

Later courses and bracing

Align units in the upper courses precisely with the units on the first course. This leaves the cutting pattern the same and aligns the cavities. It also aligns vertical joints all the way up the wall. Glue vertical joints continuously and spot glue horizontal joints every foot.

Flanges are not necessary on bucks. To get the same result as a "flanged buck," simply construct it the same way, but omit the flanges and glue it to the formwork. If you expect high wind during construction, put an occasional kicker to the wall to brace them against it. It is also a good idea to install outside corner braces at all corners. No other bracing is necessary.

Post-and-beam panel systems

Amhome, the post-and-beam panel system, involves more preplanning and less field assembly than most systems. To build with the system, you must be licensed by the company that designed it, which provides detailed instructions, specialty tools, and support. The directions below are only introductory; when building with the system, you will receive more specific ones from the company that will override those listed here. The Amhome system includes a roof structure of wooden I-joists, radiant foil, and foam that completes the superinsulation of the home. We do not cover that here; contact the manufacturer for details.

Preassembly

Amhome licensees form the panels by cutting cylinders for posts and beams out of 9¼-inch-by-4-foot-by-8-foot foam stock. Generally, 1.5-pound foam is sufficiently strong. The panels can be longer than 8 feet if your design has tall stories. Send your plans to the company, and it will return a diagram showing column placement for each panel, and drawings to show where each panel is to be set to form the walls of the house. In general, there will be one post every 4 feet, plus one at each corner and along either side of each opening. So some panels will have one cavity for a post and some will have two. There will be one beam (called the "bond beam") at the top of each story, so each panel will have a horizontal cavity along the top. In addition, vertical slots in each panel will hold 1-x-2 furring strips to serve as fastening surfaces, and horizontal channels are cut a foot or two above the bottom of each panel to hold electrical wiring.

The manufacturer provides licensed builders with special thermal cutting tools designed for making up the panels. This is usually done in advance, although custom cuts can be made in the field.

The bottom of each post cavity is reinforced with special "blowout preventers." They help hold the concrete from placing too much pressure on any weak spot that might be at the bottom of the cavity wall.

Bucks do not need flanges; because the formwork is only filled in occasional posts, there will be little outward pressure on it around openings. The sills can be solid pieces of lumber (rather than a split pair of pieces of 2-× lumber) because concrete is generally not poured below openings. The bucks will not come into contact with concrete, so codes will not generally require that they be made of pressure-treated lumber. However, depending on how exposed you expect them to be to the outside, you might prefer to use PT anyway.

Corners and setting

Form corners by abutting the two corner panels as in Fig. 5-34. Before setting them permanently, you can shim or shave them as necessary to get a precise level. However, this should rarely be necessary because the large footprint of a panel tends to span uneven spots in the foundation. Glue the panels along their bottom edges and their vertical joint. You can hold them together for drying with a long (about 6 inches) staple such as a nursery staple used for plants. Amhome's

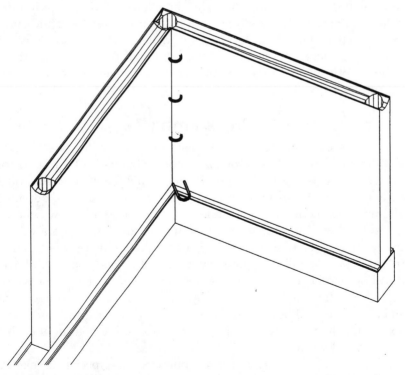

5-34 *Abutting post-and-beam panels and securing with long staples to form a corner.*

instructions for your house plan will include a sequence in which to set panels. Level all panels and glue along the bottom and the joints, as at corners. Seal all joints with an appropriate sealant.

Openings

Bucks and the panels around them are set just like flat panels (see Fig. 5-31). Set the panels alongside and below the opening, then the buck, then the panel above. Glue adjacent panels. To anchor the buck, drive galvanized spikes into the buck and through the foam to the adjacent post on either side. You can also use some other similarly long steel fastener.

Bracing

Put ladders around the top of the walls and kickers to the ladders and bucks as necessary to plumb all walls.

Floor decks

Build second-story floor decks by placing your joists on top of the finished, poured wall below. Figure 5-35 shows details. When pouring the wall below, even out the concrete along the top and insert J-bolts just as you would for a roof plate. After hardening, install top plates all around. The manufacturer specifies the use of EPS gaskets on the plates, too. Leave a few inches to either side of the rebars protruding from below, however. The concrete posts will continue along the rebar up to the story above, and the top plates should not interrupt them. Set your joists on the top plate conventionally and build the deck.

Upper stories

There are few differences to setting upper-story panels. There are no guides. Glue the panels to the ones below. Cut close-fitting notches out of the bottoms to go over the joists and plates of the floor deck.

Roof ends

Instead of concrete gable or gambrel ends, precut foam panels to the exact roof end wall dimensions. Then cut vertical slots in the faces of the foam and mount framing lumber into them. You will set these on top of the concrete walls in the field. The manufacturer can provide details.

Irregular angles and openings

For non-90-degree corners, precut the corner panel edges at the appropriate angles. Set and connect them just like the other corners. For nonrectangular openings, simply cut the desired shape out of the foam and fashion a buck to fit. Do all of the cutting in advance.

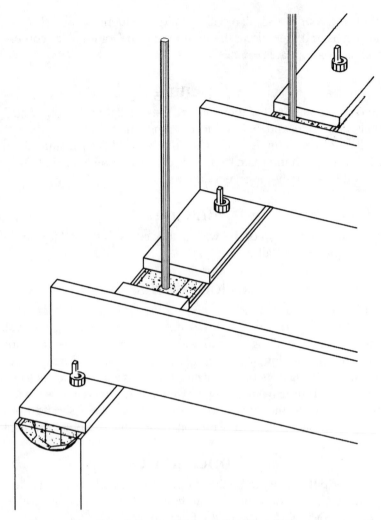

5-35 *Attachment of plates and joists for floor deck to a post-and-beam panel wall.*

Flat plank systems

The plank system units are assembled as they are set, as depicted in Fig. 5-36. Foam planks range in size from 8 inches to 12 inches high and 4 feet to 8 feet long. Ties set in their edges hold them a constant distance apart and hold them to the planks above and below. Diamond Snap-Form, Lite-Form, and QUAD-LOCK use plastic ties whose ends also serve as fastening surfaces. Polycrete uses a combination of steel rods and T-shaped plastic channel, which you interlock as in the figure. The rods hold opposing panels apart, and the channel holds adjacent panels together and serves as a fastening surface.

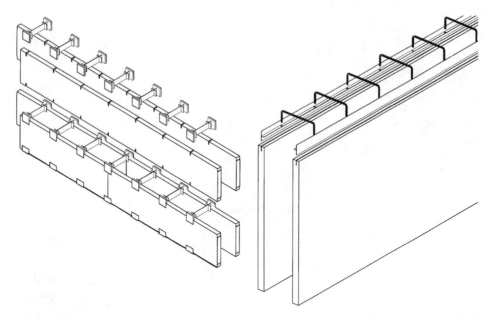

5-36 *Assembly of typical flat plank formwork (left) and Polycrete flat plank formwork (right).*

You set these systems by installing a course of planks, putting ties in their top edges, placing another course of planks over those ties, and continuing to alternate planks and ties in this fashion up the wall. This unique sequence causes a few differences in the setting procedure.

Foundation preparation

Although the other plank systems all use 2-×-4 guides, the Polycrete panels are held in alignment on the foundation with a U-channel similar to that used with R-FORMS (see Fig. 5-28), except that with Polycrete it is run continuously around the perimeter. Concrete nail it to the foundation.

Corners and setting procedure

The systems have different staggering requirements, depicted in Fig. 5-37. Lite-Form and Diamond Snap Form need not have staggered vertical joints. Thus it is easiest to set all corners and intermediate planks exactly the same on each course. At the vertical joints, bind adjacent planks across the seam with wire, as you would do with Styroform (see Fig. 5-30). QUAD-LOCK is staggered similarly to the way many block systems are: the corners alternate up the wall between "longs" and "shorts," offsetting adjacent courses by half the length of a plank (2 feet). Polycrete is also offset by half the length of a plank (4 feet), but the inside planks are also offset from the outside ones, as in the figure.

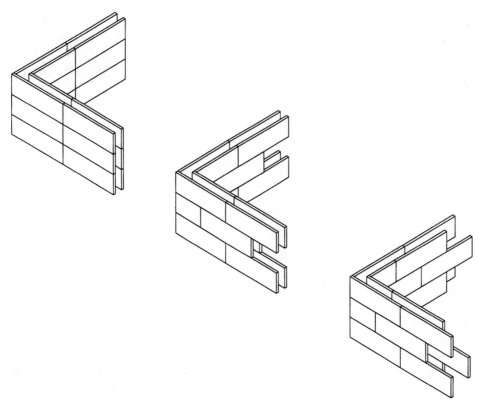

5-37 *Three patterns of flat plank setting: aligned vertical joints, vertical joints staggered between courses, and vertical joints staggered between courses and inside and outside planks.*

Diamond Snap-Form has preformed corners. Set one at each corner of the foundation. Pick a wall and set planks in between the corners from left to right. Along the bottom of each pair of planks on the first course, you will need to install "half ties." These are special ties that do not protrude beyond the edge of the foam to catch another plank. You order them from the manufacturer with your materials package. When you reach a corner, simply cut the last planks to abut the preformed corner. Do the second course exactly like the first, aligning vertical joints. When you reach the roofline, secure the top edges of the top course of planks with half ties. At the vertical joints, bind adjacent panels with wire, as in Fig. 5-36, on both the inside and outside faces.

For Lite-Form, make the first corner as done with flat panels (see Fig. 5-29): cut two planks short by the maximum thickness of the concrete, plus the thickness of one plank. These two will serve as the inside planks. Butt the four planks together to make the corner. Secure the joint with a special "corner tie," provided by the manufacturer. Along the rest of the four corner panels, place half ties on bottom and full ties on top. Work from the first corner clockwise around the perimeter. At each corner, cut the last planks to end in the corner (forming two of the four corner panels) and cut a short inner panel to begin the next wall. Secure these corners with corner ties as well. Repeat the sequence exactly on upper courses. Use half ties at the roofline, and wire adjacent panels across the vertical seams.

For Polycrete, make corners according to the same pattern as with Lite-Form. However, also cut the inner planks short so that they end at the point that is halfway down the length of their opposite outer planks. Set around the perimeter as described with Lite-Form. When starting a new wall, remember to keep the inner plank off the corner short by half the length of the outer plank. Install no ties or channel along the bottom of the first-course planks.

On the second course, reverse the pattern of vertical joints. At the first corner, cut the outer planks short to align with the ends of the inner planks on the first course. Leave the inner planks on the second course long to align with the ends of the outer planks on the first. Work around the perimeter, remembering to reverse the joints when taking off from a corner on a new wall.

With QUAD-LOCK, form "long" corners and set them at the corners of the foundation. You form these similar to the way Lite-Form corners are made, with a special arrangement of the standard ties in the corner. The manufacturer provides details. On the bottom, use half ties, and use full ties along the top. Work from left to right along the wall and cut the last pair of planks as needed to meet the next corner planks. On the second course, construct the corners the same way, but cut all planks short by two feet. Use half ties along the top course of block when you are at roofline.

Later stories

With all of these systems, end the first story with full ties or (for Polycrete) plastic T-channel. Begin the next story placing the next course of planks on these ties or channel. Set the first course of an upper story just as you would set a course on the first story.

Bracing and scaffolding

Polycrete offers a combined bracing and scaffolding system. The manufacturer has details. The others are braced just like the block systems.

Curved walls

You can curve any of these planks by making vertical dado cuts every 4 inches on their inside face and bending them, as in Fig. 5-38. You can increase or decrease the curvature by making the dadoes wider or narrower, or spacing them more closely or farther apart. Experiment for your particular curve. To begin and end the curve exactly where you want in the wall, you will probably have to make the dadoes on one end of the planks, ending somewhere in the middle.

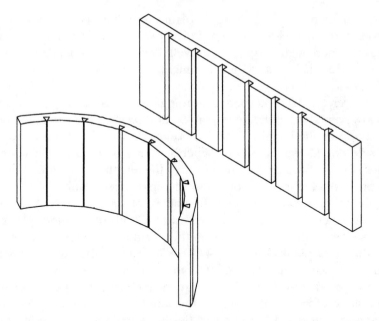

5-38 *Curving a plank by making dado cuts (upper right),
then bending and gluing (lower left).*

Put glue in the dadoes before bending the planks to hold the curve and keep
the planks strong along the joints.

   With Lite-Form and Diamond Snap Form, it is also possible to stand the
planks on their ends and bevel cut the long joints to approximate a curve, much
as is done with panels (see Fig. 5-32). This arrangement will create a vertical
joint between the curved panels (set up lengthwise) and the straight sections of
wall (made of panels set on their sides). You can tie these together with full ties,
as in Fig. 5-39.

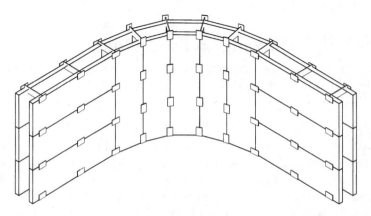

5-39 *Forming a curved wall section by bevel cutting the
edges of flat planks and setting them vertically.*

6

Placing concrete

Measure thrice and place once. ICF formwork is easy to correct before it is filled. It becomes harder afterward. If you have not built an ICF wall before, do all the finishing touches on the formwork before you schedule the pour. Then leave at least one free day in between for inspection and to double-check everything at the site. Once you are more experienced, you can devise your own, tighter schedule. (Note that here, instead of "pour concrete" we use the technically more accurate term "place concrete.")

Finishing touches and preparation

Form a sleeve (depicted in Fig. 6-1) for things that must run through the wall—wiring, vent pipes, water lines, and so on. The sleeve keeps a clear path through the formwork. When using a flat panel, flat block, or grid block system with fastening surfaces, it is best to make a sleeve for each planned penetration. You can core drill holes through the wall after it is filled with concrete, but the sleeve is usually cleaner and easier.

If you are using a grid-panel or grid-block system without fastening surfaces, you generally do not need sleeves for penetrations of 1-inch diameter or narrower. You can cut easily through the foam ties later. However, you will be constrained to locating the penetration in one of the ties. If you are using a post-and-beam system, you generally need no sleeves at all. You can simply cut through unfilled portions of the wall later. The one exception is when filling post-and-beam block completely (as for a basement). In that case you will need sleeves for large penetrations; small ones can go through the ties.

Table 6-1 includes the most common penetrations (under "Insertions in the wall") for you to check against. To make a sleeve, get a piece of PVC pipe just larger than the item that is to go through (3½-inch inside-diameter pipe for a 3-inch round vent; ¾-inch pipe for an ordinary electrical cable, and so on). Cut it to a length equal to the total width of the ICF units you are using (see Table 3-1). Mark the desired position of the penetration on the wall, tracing around the end of the pipe. Make certain that the wall has no rebar or other obstacles in that spot. Also, avoid spots that would require you to cut ties. This weakens the

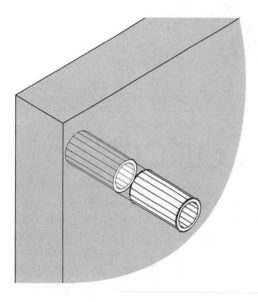

6-1
Insertion of a sleeve into formwork.

formwork. Cut the marked circle out of the foam with a sheetrock or keyhole saw. You will need to make two cuts, one on each face of the formwork, to go completely through it. Slide the pipe into the holes flush with the wall and glue it into place.

Table 6-1. Prepour Checklist.

Item to Check

Inspections

> The building department has inspected and approved the steel reinforcing bar (if required).

Formwork

> The footprint matches plans in layout and all dimensions.
> The top extends exactly to the roofline level (if top story) or near top of floor joist level (if an intermediate story).
> The top is level all around (if top story).
> All openings in the plans are in the wall.
> All openings have a complete buck.
> All openings are in the location and of the dimensions planned.
> All walls measure plumb.
> All corners measure square.

Lintel block (for post-and-beam block systems only)

> There is a course of lintel block around the entire perimeter at each scheduled beam level.
> There is a course of lintel block at the level of the floor deck on each joist-end wall.
> All lintel block is open in the bottom along all vertical cavities to be filled with concrete, and stopped up above all other vertical cavities.

Item to Check

Rebar and cavities

All horizontal rebar is in place, including:
at regular planned intervals;
above each opening (except for post-and-beam systems);
Each wide opening has extra horizontal rebar above.
All horizontal rebar laps the adjacent bars sufficiently.
All rebar over an opening extends sufficiently to either side.
All vertical rebar is in place, including:
at regular planned intervals;
at each corner;
along either side of each opening.
All vertical rebar is in the collar at bottom.
All vertical rebar is tied at top to center it in the cavity.
All vertical rebar sticks up sufficiently above the formwork (except for rooflines).

Insertions in the wall

All embedded fasteners are in place, including:
duplex nails or J-bolts in the sides of each buck.
duplex nails or J-blots where the end studs of interior walls will attach (optional).
U-wire along the exterior (for PC stucco siding on a system without fastening surfaces).
Brick ties along the exterior (for a brick veneer on a system without fastening surfaces).
Duplex nails where exterior furring strips will attach (for nailed and screwed sidings on a system without fastening surfaces).
All sleeves are in place, including for:
electrical, phone, CATV, water, and sewage service.
any other service from outside.
wiring to outside electrical fixtures.
wiring and pipes to outside AC units.
pipes to outside spigots.
bathroom, dryer, stove, and air exchange vents.

Floor deck (if any)

All ledger boards (if used) are complete, including:
cutouts behind the boards in each cavity.
J-bolts into the cavities.
All pockets (if used) are complete, including:
one pocket blocked out at each joist location (precisely).
all blocks protruding sufficient depth into the cavities.
all blocks oiled on the cavity end and glued into place on the interior end.

Roofline (if top story)

The formwork is cut precisely to roofline height at all points a roof will be attached.
A mark shows the position of each metal strap or J-bolt required.
Strap marks fall on the exact position of each rafter or truss.
J-bolt marks fall between rafter or truss positions.
Enough metal straps or J-bolts are on site.

Table 6-1. Continued.

Item to Check

Gable or gambrel ends (if any)

    The lines of the roof ends match the designed angles and elevations.

    The elevation of the roof ends allows for a top plate.

    A course of lintel block tops the roof ends (post-and-beam systems).

    Top plates are in position on the roofline, with gaps between adjacent plates and blocks nearby to fill the gaps.

Bracing

    All guides are firmly in place (if first story).

    All corner braces are firmly in place.

    Exterior and interior corner braces are wired together at regular intervals (systems with cut or assembled corners).

    Plywood covers cuts at the right-hand side of each wall (block systems).

    2-x-4s or ladders brace the top of wall at all points.

    There are kickers to all bucks, corner braces, and at intervals along walls with no openings.

    There are intermediate vertical braces (if required).

Form Soundness

    There are no gaps in the walls that concrete might get through.

    All weak spots are glued or braced, including:

        Gaps around sleeves, pocket blocks, and other insertions.

        Cut or irregular joints between ICF units.

        Wall areas where ties are broken or missing.

        Spots where the foam faces have been cut, damaged, or punctured.

    All joints between units have been glued or secured with manufacturer-supplied connectors (except for friction-fit blocks).

Miscellaneous

    There is tape covering the top edges of the formwork (intermediate stories only).

    All components and materials below that need to remain clean are covered.

    A complete blowout kit is on-site.

If the wall you are pouring is an intermediate story, cover the top edges of the formwork with a removable tape. If you do not, bits of concrete will get on them and make setting the next course difficult. You can clean it off later, but prevention is usually faster.

If the wall instead goes to the roofline, you will need to place J-bolts or metal straps in it to anchor the roofing lumber. In high-wind areas, the building department often requires hurricane straps that connect directly to the rafters or trusses. Otherwise, most carpentry crews prefer to have J-bolts so that they can bolt top plates to them. This gives them more freedom in positioning the roofing members.

Mark the position of each strap or bolt clearly on the formwork. If using straps, put one mark at exactly the position of each roofing member. If using bolts, locate them between the expected rafter/truss positions.

Cover anything below the top of the formwork that might be damaged by concrete. Some will certainly fall. The force of the fall might damage some products, and dried concrete might be hard to remove from others. Things like window and door frames (if any are set yet), bulkheads, and piles of material can be quickly covered with tape or tarps.

Finally, make a "blowout kit." Breaks in the formwork during the pour are rare and get rarer with experience. But it is best to have the hardware ready to handle them quickly just in case. If your system has fastening surfaces, cut six squares of plywood. The plywood should be at least ½ inch, and the squares should measure about 2½ feet on a side. Also have wallboard screws and a screw gun on hand. That is the entire kit.

If your system has no fastening surfaces, drill one ⅜-inch hole at each corner of each plywood square, about 2 inches off the corner. You will also need 12 pieces of ¼-inch threaded rod and 24 washers and nuts for them. The rods should be at least 4 inches longer than the Total Width of the ICF units (see Table 3-1). In the case of a blowout, you will hold the plywood to the wall with these rods instead of by screwing it to the wall.

Ordering inspections, equipment, and materials

Some building departments require an inspection of the reinforcing steel before the pour. They will verify that the sizes of rebar being used, the number of bars, and the positioning of the bars match the plans and code. Check with your building department far in advance to find out whether the department requires such an inspection. If so, schedule it to come as soon as practical after all the steel is in place.

Most equipment requires at least two days' advance notice for rentals, and most ready-mixed concrete suppliers require one. Check with your local companies.

If you are filling walls below grade, you can use the chute that comes with the ready-mix truck to place the concrete. No pump rental is necessary. Your site will have to be graded so that the truck can get to points all around the perimeter about every 20 feet. You will also have to keep the driver placing at a pace slower than usual.

For above-grade walls, or if you prefer on below-grade walls, use a line pump or a boom pump. If you have not built with ICFs before, order your pump with a 2-inch line if that is available. The pour will go more slowly, and you will have to pay a little more for a high-flow concrete mix. However, you will have to go slowly anyway because the tasks of the pour will be unfamiliar to you. And with the narrower line, there will be less chance of leaving a void or having a blowout. If using a boom pump, ask for the hose arrangement described in chapter 2 (see Fig. 2-7). If only 3-inch lines are available, you can use one, but you will need to monitor the placing more carefully.

One story of exterior walls of an average size house can be filled in half a day, so order the pump for that long. For a large pour you may want to reserve it for the whole day. As you gain experience these times will drop. Schedule the concrete to arrive about half an hour after the pump. (The pump requires setup time.)

You will probably need to have two concrete deliveries. Most trucks hold 10 yards of concrete, and an average pour requires between 10 and 20 (although post-and-beam systems require less). Ask the concrete supplier to deliver in two equal loads about 1½ hours apart. This will keep you supplied but minimize the amount of time each truck must wait, avoiding time charges from the supplier. Most suppliers allow 5–7 minues per yard of concrete. If you hold a truck longer, you get a surcharge.

The amount of concrete to order depends on the size of the walls and the system you are using. Some ICFs have larger cavities and therefore hold more concrete. To figure your needs:

1. Count the number of units you will be filling. Where you have made cuts, there will be half-units and quarter-units. Add these fractions in, too. Make sure you add the numbers for all walls together to get the total for the entire story.
2. Find the cubic yards per unit for your ICF system in Table 3-1. If you are using a post-and-beam system, bear in mind that this number will vary depending on how many cavities you fill.
3. Multiply the total number of units by the yards per unit.
4. For waste, increase the yards calculated in step 3 by 1 yard or 5 percent, whichever is greater.

Most ICF manufacturers have recommended concrete specifications to use with their forms. You should always follow the manufacturer's directions or those of an engineer. Some systems, for example, require high-strength concrete or other unusual features not in the typical specifications. Following are some general guidelines that are meant for initial planning purposes only.

As a general rule, if you are placing with a chute and you do not need an unusually high flow to the concrete, you can use the following mix:

- 3000 psi
- ⅜-inch aggregate
- 6-inch slump

This is standard residential construction concrete with a slightly smaller stone and a higher slump.

If you are using a pump with a 3-inch line or a system with narrow cavities (including most grid and all post-and-beam systems), it is safer to use a concrete with greater flow. You can ask for:

- Pump mix
- ⅜-inch aggregate

Most suppliers will automatically use ⅜-inch aggregate when they hear "pump mix," but it is worthwhile being sure.

If you are using a pump with a 2-inch line, you will need a concrete with an even greater flow. You can explain the situation to your ready-mixed concrete company and have them choose the ingredients. As you gain experience, you can specify individual ingredients yourself to achieve a balance of high flow with low cost that meets your needs. Generally, it will contain less aggregate and more of the other dry ingredients than standard concrete mixes.

Especially if you need a slump higher than 6 inches, consider asking the concrete supplier to use a plasticizer to achieve the slump. There is some additional cost, but the concrete will be stronger and more water-tight than if the slump is achieved completely with water.

Double-checking

After finishing the formwork but before the pour, check the formwork, bracing, and site for potential problems. If you have set aside the day before the pour to do this, you will have time to correct anything serious or delay the pour.

Table 6-1 lists some of the major items to check. However, do not limit your inspection to these. Look for anything that might cause difficulty during the pour or later construction or occupancy.

If anything is wrong with the formwork, you can cut out sections (as for missed openings, too-small openings, or too-tall walls), fill in sections (unnecessary openings, too-large openings, too-short walls), or cut out and replace sections (incorrect wall placement, wrong type of block). Filling in can leave long, continuous joints that form weak spots. Glue these continuously, then reinforce them as described in Table 6-2.

It is easy to move vertical rebar by pulling it out and resetting from the top of the wall. To move horizontal bar, slide it in and out through holes in the nearest corner, as in Fig. 6-2. You might also have to cut holes in the formwork to get to the bar to move it. When finished, fill up all holes—small ones with adhesive foam and large ones with cutouts of foam glued in place. Reinforce this patching with one of the methods listed in Table 6-2.

If you have left embedded fasteners out of the sides of a buck, you can drive ring nails through them at regular intervals. The rings will anchor into the concrete. This saves having to remove the bucks. For post-and-beam systems, drill holes and slide 9-inch pieces of rebar through to the nearest adjacent posts.

If you have unnecessary embedded fasteners in the foam, you can pull them through and patch the hole. If you have left out some, you can push them in from the outside and fill any gaps you have created around them.

Unnecessary sleeves can be cut out and the holes patched. Missed ones can be put in at any point before the pour. Mistakes in the pockets or ledgers for a floor deck can be ripped out and recut. If they are serious, consider removing the affected sections of formwork completely and replacing them. Bracing can be added at any point before the pour. It should be possible to adjust the plumb

Table 6-2. Procedures for Reinforcing Seams and Weak Spots.

Type of System	Procedure
Flat panel	
Styroform	Wire looped around tie ends across seam (see Fig. 5-30), or strapping or plywood placed flat over the seam and screwed to the tie ends.
R-FORMS	R-FORMS bracing, or strapping or plywood placed flat over the seam and screwed to the tie ends.
Grid panel	Vertical braces over the seam, or plywood placed over the seam and held flat by vertical braces.
Post-and-beam panel	Vertical braces over the seam, or plywood placed over the seam and held flat by vertical braces, or (over portions of the wall to go unfilled) nothing.
Flat block	Strapping or plywood placed flat over the seam and screwed to the tie ends.
Grid block with fastening surfaces	Strapping or plywood placed flat over the seam and screwed to the tie ends.
Grid block without fastening surfaces	Vertical braces over the seam, or plywood placed over the seam and held flat by vertical braces.
Post-and-beam block	Vertical braces over the seam, or plywood placed over the seam and held flat by vertical braces, or (over portions of the wall to go unfilled) nothing.

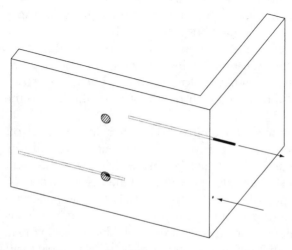

6-2 *Removal and repositioning of horizontal rebar.*

of the walls merely by shifting kickers. For any large weak spots or seams you have not already reinforced, use the methods listed in Table 6-2. For small ones, some extra adhesive foam is usually sufficient.

Finally, remember to dress properly. Since concrete can sometimes irritate skin, it is now recommended that you cover up: boots, long pants, shirt with long sleeves, gloves, hard hat, and eye protection.

The pour

Note that many system manufacturers have their own recommendations for placing concrete that might differ from the ones in this chapter. Follow the manufacturer's instructions if they are available. The following instructions are only for planning purposes and to serve as a guide if no manufacturer instructions are available.

While the pump is setting up, triple-check anything that is questionable and organize the crew. Three or four workers make an ideal pour crew. If you have three, put two up on the scaffolding and one on the ground/floor. One of the up crew can come down if needed there. If you have four, put two up and two down.

To place concrete, the crew will start at a corner and work around the perimeter in a clockwise direction. Pick any corner. The sequence of placing depends on whether you use a post-and-beam system or not.

Flat and grid systems

Begin with the wall to the right of the starting corner. First place concrete through each window sill. (You can skip small ones, such as basement windows.) Start with the leftmost opening. If the opening is narrow, place the concrete in the middle (halfway between the jambs) until the peak of the concrete comes up to the bottom of the buck sill or is 3 feet high, whichever is lower. If the opening is wide, place concrete once every 2 feet. As before, allow the concrete to peak at the bottom of the sill or up 3 feet.

Go back to the starting corner and place concrete from 2 feet to the right of the corner. Fill the formwork there up to a height of 3 feet. To verify the height of the fill, the down crew can hit the formwork with the flat faces of short 2-x-4 blocks. Where it is full, they will hear a dull thud and feel a more solid wall.

Both up and down crews should be alert for voids—large air pockets created because the concrete got hung up at some point. The up crew can often spot these because the concrete suddenly appears very high in the cavities. They can correct it by sliding a long piece of strapping or rebar up and down in the stuck concrete to dislodge it. The down crew should be checking constantly by hitting the wall in the area of the pour at all heights. If they start hearing a hollow sound, they can tap at higher points until they find the stuck concrete above. They can dislodge it by placing the 2-x-4 block flat against the wall and hitting it with a hammer to shake things free. Position the block and hit it over the end of a tie.

The down crew should regularly do light tapping of the wall with 2-x-4 blocks and hammers throughout the area being filled even if there are no voids. This consolidates the concrete and helps it fill even small crevices. Once the concrete is up to 3 feet without voids, the up crew moves the hose to the right 4 feet and repeats the procedure. Continue to the end of the wall in this way, stopping with the hose or chute positioned 2 feet before the right corner.

Note that some ICF manufacturers recommend power vibrating equipment to consolidate the concrete, not just tapping. Others do not because of the possibility that it will help the concrete work into weak spots and hasten blowouts. If you wish to use it, consult your manufacturer's literature or the manufacturer directly.

When placing concrete alongside window openings, the concrete should work its way underneath the buck. If the buck sill is below 3 feet high, you want the concrete to rise at least up to the bottom of the sill all along it, but not above the top of the sill. You will place along the left side first. While the up crew is pouring, the down crew taps the wall as necessary to get concrete to migrate over and rise up to the sill. If some pushes up through, they must even it off flush with the top of the sill or scoop off the excess. Protruding concrete would make it difficult to set the window and finish trim later. The procedure is the same on the right side. Do not yet place concrete above any door or window lintels that are over 3 feet up.

After the first wall is finished, pour the second in the same way. Then continue around the perimeter until you return to the starting corner. If less than an hour has passed since beginning the pour, wait until that time before proceeding to the second layer. This allows the poured concrete to stiffen, stabilizing the formwork and reducing the pressure at the bottom of the wall.

However, in general avoid delays in placing concrete. If you allow any section of concrete to harden completely before placing more on top of it, a "cold joint" (weak spot) will form. In addition, holding the concrete truck over 5–7 minutes per yard (depending on your supplier's policies) results in surcharges.

Begin placing again along the starting wall. If the story you are filling is 9 feet high or less, fill up to the top in this pass. If it is higher, place only an additional 4-foot-high layer.

Place as you did the first pass. If there are any window openings with sills more than 3 feet high, fill up to the sills, as described previously, before doing the rest of the wall. Place over door and window buck lintels as necessary to fill up to the recommended height.

If a third pass is necessary, do it as you did the second.

Post-and-beam systems

When filling post-and-beam systems, you fill occasional cylinders instead of a continuous wall. The concrete must also flow sideways where there are beams to be filled in the middle of the wall. These are narrow channels, so some tapping may be necessary. But otherwise there is little need to get the concrete to "spread out" as with other systems.

Figure 6-3 shows a typical pattern of concrete in a post-and-beam wall. The pour always involves filling vertical posts at regular intervals (usually 2–3 feet), at the corners, and along either side of openings, plus a horizontal beam along the top. Depending on the house design, the local structural requirements, and the manufacturer's recommendations, you might also need to fill one or more beams farther down. And if possible, fill some vertical posts below each window opening to support the window. Fill one post near each end of the opening, plus one about every 3 feet in between. Avoid leaving sills unsupported for distances over 3 feet.

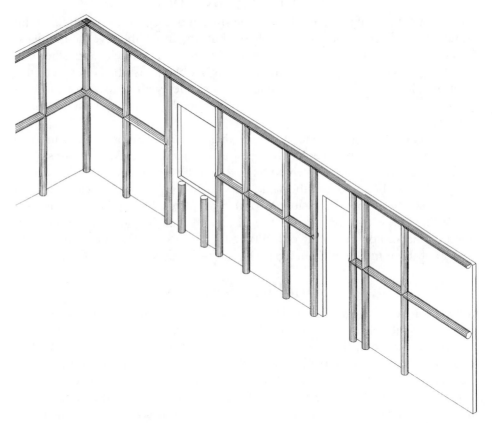

6-3 *A typical concrete pattern for a post-and-beam wall.*

To begin placing, start at a corner post. Aim the concrete at the edge of the post cavity so that it trickles down the side. The down crew can tap as usual to verify that the concrete is filling completely. If there are to be horizontal beams part way up the wall, the crew must also tap to move concrete sideways. They continue this until concrete flows through to the next posts on either side (both to the left and the right).

When the first corner post is filled and concrete is piling up in the lintel block over it, move to the next post and repeat. Also fill the beam along the top of the wall in between the two posts. When filling a beam in the middle of the wall, make sure that the concrete flows all the way back to meet the concrete flowing in the beam from the left, and to the next post to the right.

Continue around the wall. Before beginning the next wall, fill the posts under the window bucks. Then fill the next wall to the right with the same procedure, and continue clockwise around the perimeter until all walls are filled.

Completing the pour

With most systems, if you are filling an intermediate story, you need to fill up far enough to back the floor deck completely but not above the level of the deck. Otherwise you might have concrete interfering with door bucks and other features of the second story. One of the up crew should adjust the concrete to this level by following the pouring around the perimeter with a trowel or piece of strapping. Even out the concrete to the correct height, and shove any excess to the right.

If you are placing concrete at the roofline, or if you are using the Amhome system (for either an intermediate or top story), the trailing crew member should flatten the concrete exactly, even with the top of the formwork. After the second wall is filled and evened out, the same crew member should go back to the first wall and set the embedded fasteners for the top plates. Doing this after the next wall is filled gives the concrete time to stiffen. Push each strap or J-bolt into position, as marked on the formwork. Where this makes the concrete uneven with the top of the form, even it out again. Figure 6-4 shows the results.

Handling blowouts

Occasionally concrete will push through a weak spot in the formwork, tear the foam, and spill out. The first line of defense is to prevent such blowouts.

During the pour, the down crew should regularly sight down the wall to look for bulges. If one appears, they call to the up crew to move the pour farther down the wall temporarily. They then take one of the plywood squares from the blowout kit and place it over the bulge. If the system you are using has fastening surfaces, the crew can secure the plywood by screwing it to the formwork. Otherwise, they can brace it flat against the wall with a kicker, as in Fig. 6-5. Some bulges will flatten out by themselves. If they do so before the down crew can get the bracing in place, the crew can decide to skip it. After bracing, the up crew go back to fill the area with the bulge to the appropriate level.

If the formwork does blow out, the up crew should immediately move to fill the next wall. They can continue there. The down crew will shovel up the spilled concrete and remove concrete from inside the formwork through the blowout hole until it is just below the bottom of the hole.

6-4 *Hurricane straps embedded in the top of a wall.*

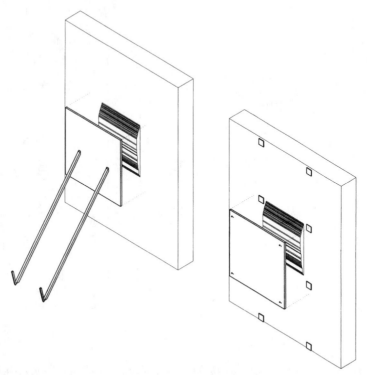

6-5 *Bracing a bulge in formwork without fastening surfaces (left) and with fastening surfaces (right).*

The down crew then glues the blown out piece of foam back into place and secures the formwork in that area on both sides with the blowout kit. To secure the formwork, they simply screw the plywood squares to both sides of the formwork (if there are fastening surfaces), or hold the squares flat to either side with threaded rods run through the foam. Figure 6-6 shows details. Support on the back side as well as the blown side is necessary because each face of an ICF unit is supported by the opposite one, through the ties. With the blown face damaged, it will no longer support the opposite face adequately.

When pushing threaded rod through formwork without fastening surfaces, try to put them through the foam ties or (if you are using a post-and-beam system) unfilled cavities. This will allow you to pull them back out later, as they will not be locked into the concrete. At their next convenient break point after the blowout is repaired, the up crew should return to finish filling the wall on which the blowout occurred.

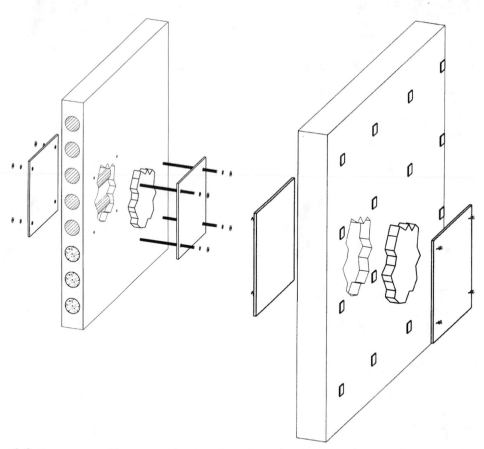

6-6 *Repairing a blowout in formwork without fastening surfaces (left) and with fastening surfaces (right).*

Cleanup

A few steps of cleanup during and after the pour will save work later. Bits of splattered concrete can accumulate on the floor, ground, and anything else lying below. It is easiest to clean up about 1–2 hours after the concrete has fallen, when it will be stiff enough to come up in clumps but not yet hard enough to adhere. The down crew should keep track of spills and scrape them up at about the right time.

After the pour is completely done and the concrete finished, remove tape put over the top edges of the formwork. If any concrete landed on the foam edges somehow, scrape it off with a putty knife. If you wait until it dries instead, it will adhere to the foam and the work will take longer.

Twenty-four hours after the pour you can remove bracing. Remember to leave up the corner braces if you filled an intermediate story and your system has no fastening surfaces. The braces below will support the ones on the story above. Also remove the plywood squares of any bulge or blowout repairs. If any of the threaded rods used for a blowout repair are embedded in the concrete, cut them flush with a hacksaw.

Weather considerations

Few weather conditions affect a pour because the foam formwork insulates the concrete, allowing it to cure almost regardless of outside temperature or humidity. It is usually acceptable to pour down to temperatures of 10 degrees Fahrenheit. If the temperature is below freezing, or will go below freezing in the next couple of days after the pour, insulate the top of the wall after pouring. To do this, lay fiberglass batts or some other type of insulating blanket over the concrete everywhere it is exposed. This prevents the surface from freezing, and freezing prevents the concrete from curing properly. Remember that fiberglass and many other insulations lose their insulating ability if they are wet, so if rain or snow is likely you will need to cover the insulation with plastic sheet or use a moisture-resistant insulation instead.

In extremely hot, dry weather, concrete can dry out, also preventing good curing. This is rarely a problem in ICF formwork because foam is a good moisture barrier and so little concrete is exposed. But as a precaution, put plastic sheet over the exposed concrete at the top in dry weather. Remove it after the concrete hardens and before continuing construction.

Light rain is not a deterrent to placing, either. However, do not place in a hard rain. If rainwater accumulates in the concrete, it will cause voids and weak spots.

It is also acceptable to place into wet formwork. Any accumulations of water usually leak out quickly, but if there is any standing water in the cavities, drain them before placing.

7

Remaining rough carpentry

There are several types of framing that attach to the ICF walls. If you are building multiple stories out of your ICF, you build the floor decks between erecting the stories, just as with frame construction. Any exterior walls built of frame must also be added as you work up the structure. After topping off the exterior walls, you build the roof. Finally, various interior walls will abut the ICF walls, and you will have to decide how to attach these.

Floor decks and roofs

You can normally start setting a floor deck within 48 hours after filling the story below. However, consult the manufacturer or an engineer to be sure. In the meantime you can remove bracing from the wall.

To build a pocketed floor, knock out the pocket blocks. Then set the joists in the pockets. Add shims if necessary to level the joists. You might also want to add shims to the sides of the pockets to keep the joists steady during attachment of the decking.

Some building departments require a "fire cut" on the pocket end of each joist. These are angled cuts off the top corner of the joist to allow the joists to fall out of the wall without putting stress on the wall during a fire or other disaster. Figure 7-1 depicts fire cuts. If these are required, simply make the cuts before setting the joists.

Remember to fill the gaps in the formwork around the joists where they enter the pockets. Otherwise, concrete might leak out during the pour of the next floor. If you are using a ledger, attach joists with hangers as you would on any hung floor.

Instructions for the roof are similar. You can usually begin within two days after the pour of the top story, but follow the recommendation of the manufacturer or an engineer. If you are using a top plate, attach the roofing members as you would onto a conventional frame house. If you intend to set rafters directly

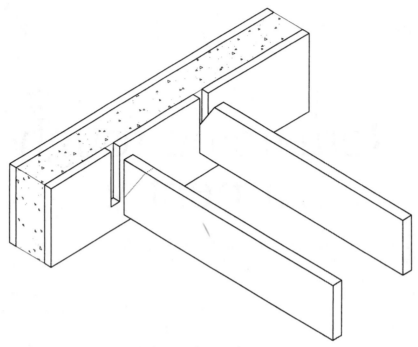

7-1 *Fire cutting joists before pocketing them.*

on the plate, you might need to make some allowance to fit over the thick ICF wall below, as in Fig. 7-2.

If you are using hurricane straps, the rafters or trusses will rest directly on the ICF wall, as they would on the top of a frame wall. However, you will attach them by nailing on the strap, rather than nailing to the wall.

Exterior frame walls

If you build a section of the exterior wall out of frame, you should begin with a "buck" around the perimeter of the section, as in Fig. 7-3. Build the buck of 2-x lumber as wide as the total width of the ICF units (see Fig. 3-1). Fill in with conventional frame. The framing inside the buck can be whatever width you prefer. However, if you use a lumber width less than the ICF unit width, you will have a recess in the wall to finish.

Attach and build around the wall buck as you would a door buck. Outfit the jambs with J-bolts to anchor into the concrete. Add kickers as necessary to hold the buck straight against the weight of the concrete during the pour. If the frame wall will rest on a cantilevered floor, build the floor first. Then build the buck and framing, as in Fig. 7-4.

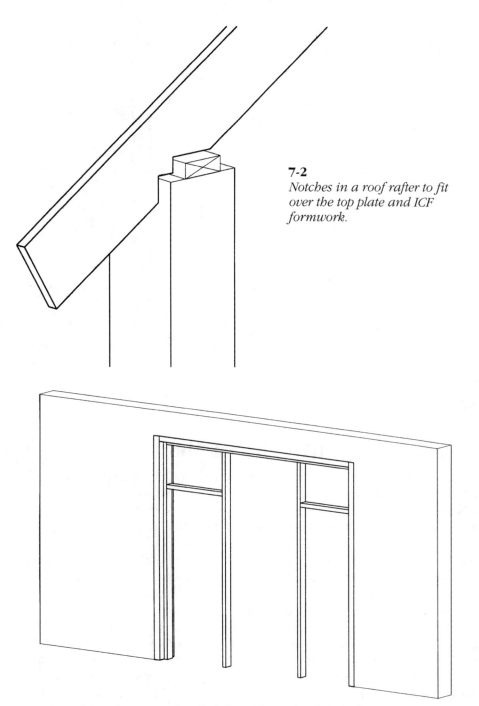

7-2
Notches in a roof rafter to fit over the top plate and ICF formwork.

7-3 *A framed section of wall: full-width 2-× buck filled with 2-×-4 framing.*

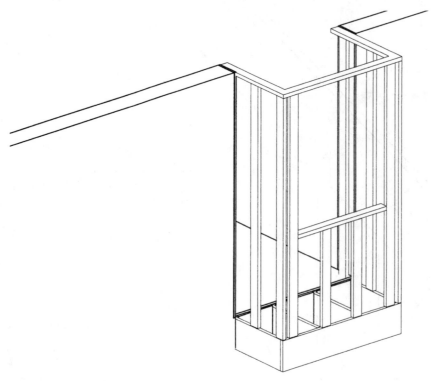

7-4 *A framed, cantilevered bay.*

Interior walls

Interior walls that butt into ICF walls can be attached by the end stud in several different ways. If you have anchored duplex nails into the concrete at that point, it is easiest to pound scraps of 2-x-4 onto the nails and then nail the end stud to them, as in Fig. 7-5. If you do this, you must remember to build the interior wall 1¾ inches short of its design length to compensate for the 2-x-4 on the end. If you have embedded J-bolts, set the interior wall temporarily in position and hit it to the bolts. Then drill where the bolts dented the stud, thread the stud over the bolts, and screw on washers and nuts. Drill the holes a little large; you might need some play to set the wall to plumb.

If you did not embed fasteners into the concrete, you have several other options. If your ICF system has fastening surfaces and some of them fall at the intended location of the end stud, you can screw or nail to them. This provides a medium-duty connection. If fastening surfaces are not available, you can glue the end stud to the foam, nail it with a concrete nail through the foam and into the concrete, or both, depending on how strong a connection you need. If you use glue without nailing, you will need to hold the stud tight against the foam somehow while the glue dries. If you use concrete nails, they must be long enough to go well through the foam.

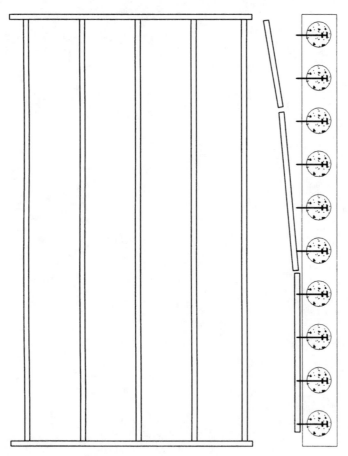

7-5 *Attaching an interior wall by first putting wooden blocks onto protruding duplex nails.*

8

Installing utilities

ICF walls can hold just about any wiring, piping, and venting that frame walls can. However, the installation procedures are different.

Penetrations

As discussed in earlier chapters, there are a few ways to run something from the inside of an ICF wall to the outside. One that always works is the preinstalled sleeve (described in chapter 6). This plastic pipe embedded in the wall allows the trades to push their own wires, pipes, or vents through whenever they do their rough-in. You must make sure that the sleeves are not covered over on both sides by the wallboard and siding crews, or the sleeves can be difficult to locate again. Once the line or vent through is installed, the gaps between it and the sleeve must be sealed outside to prevent air and water from coming in. Different crews have different ways of handling this. One that generally works is to fill the gaps with a heavy-duty sealant and have the siding crew cover over.

If you need to run penetrations through parts of the wall that have no sleeve installed, there are a few options. If you used a post-and-beam system and it is only partly filled with concrete, you can usually get any penetration you want through one of the unfilled sections of the wall. Simply cut out the foam with a wallboard or keyhole saw to make a hole of the appropriate size and shape. Then thread the line or vent through and seal the gaps. If you used a grid panel system or a grid block system without fastening surfaces, there are foam ties that can be cut out in similar fashion. With any of these systems you can cut a 1-inch-diameter hole, and with some the tie is thick enough to cut out more than 2 inches in diameter. If you used a type of system without any such foam pathways, you will need to cut through concrete. Core drill with a concrete-cutting hole saw of the proper diameter. Many of the trades carry their own. Cut where there is no rebar, as cutting a bar weakens the wall.

Surface mounting electrical lines

To install electrical wiring and boxes, cut out recesses in the foam surface of the wall to hold them. Figure 8-1 shows some typical cuts.

Begin by marking the location of each box on the foam. If possible, trace around an actual box. This will give the exact size for precise cutting later.

When you have a choice, locate boxes where the foam is thickest. In spots where the foam is thin, you might need to use shallow boxes. Check the ICF units you are using to be sure about this. With some flat ICF systems, you will need all shallow boxes. With grid systems you can use standard boxes if you locate them where the foam is deep. For grid panel systems and grid blocks without fastening surfaces, that means putting the boxes over a tie. For grid blocks with fastening surfaces, you would put the box immediately alongside a tie. At other points on a grid wall, you can put shallow boxes. With most post-and-beam systems you can mount a standard box on any un-filled part of the wall, and mount a shallow box directly over a filled post or beam.

Next draw lines where you want to run the cable. For almost all systems, cable can run anywhere. The main exception is systems with steel ties. When using one of those, do not run across a tie, or you will have to cut steel later. It is easy to go around steel ties; they are all designed to run only part way up the ICF unit, leaving open horizontal paths as well as the vertical ones.

8-1 *Some chases and cutouts for electrical cable and boxes.*

Now cut recesses where your box marks are, and cut chases for the cable lines.

Ideally, cut the box recesses to exactly the depth at which you want to set the boxes. You can bend a hot knife blade or set a router bit to do this. It is harder with an ordinary knife. Cut as close to the lines as possible so that the box slides in without much force, yet the foam holds it in place. If this is too difficult, cut a little wide. Set the depth of the chases to at least the minimum setback for cable required by your local code. In most areas a chase of 1½ inches deep is adequate. The width of the chase is ideally the same as the cable width (so the cable snaps into the chase), but a little wider will do.

If you cut the chases with a hot knife, you might want to save the long, thin cutouts to put back later. Tack each one onto the wall nearby with a nail, as in Fig. 8-2. This will keep the cutouts handy.

Set the boxes in place. With most of the block systems that have plastic or metal ties, a box positioned next to a tie can fasten directly to the tie as you would fasten to a stud. Check the manufacturer's literature first to verify that the

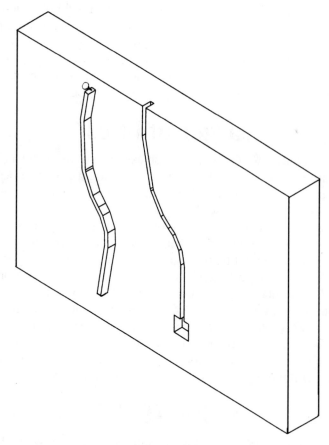

8-2 *Tacking the foam from a chase onto the wall for later replacement.*

tie is designed to do this. Remember that metal ties require a screw, not a nail. You can also glue the back and/or sides of the box for extra strength, if you wish. With other systems, you can usually just glue the box. However, some building department officials will not accept glue alone. If that is the case, you can concrete nail through the back of the box to the concrete.

Finally, pull the cables. Pull each cable end all the way to the box, then go back and press the cable into its chase. If necessary, you can hold the cable in place with a squirt of adhesive foam every 2 feet.

If you wish to restore insulation to the chase, you can run a continuous bead of foam over the cable. If you used a hot knife to cut the chases and saved the cutouts, you can press them back in instead.

Surface mounting plumbing lines

You can install small pipes on the wall surface by cutting chases into the foam, much like with electrical cable. Check the thickness of your system's foam faces to determine how big a pipe will fit. Almost every system will hold one-inch-diameter pipe, and most are deep enough to hold two-inch in at least some places.

After cutting the foam away and setting the pipe, you can fasten to plastic or metal ties (if your system has them), or to the concrete with concrete nails, as in Fig. 8-3. If there is extra space, you can restore some of the insulation by running a bead of adhesive foam in the chase over the pipe.

Larger utilities

To install larger utility lines (such as vent stacks, drains, or ducts) along exterior walls, there are several alternatives. If you are using a post-and-beam system, you can cut a large chase, just like the smaller ones described previously, so long as the pipe or duct does not have to cross a concrete-filled post or beam. You can, of course, build a frame chase on the inside wall surface to house the line. You can also reroute the line through interior walls instead.

If you must run a large line through a concrete-filled section of an exterior wall, you can plan ahead and insert the pipe inside the cavities before the pour. Consult the ICF manufacturer or an engineer before doing this: It will weaken the final wall. Cut away the inside surface of foam to allow the line to fit partly into the foam and displace as little concrete as possible. Figure 8-4 shows details. Be sure to glue the foam closed and reinforce the cut section after setting the line to eliminate any weak spots the wet concrete might seep through. The pour will surround the pipe with concrete, locking it into place.

Sizing HVAC

Because of the high R-value and low air infiltration of ICF walls, homes built with them require smaller furnaces and smaller AC compressors than conven-

8-3 *Pipes fixed to the wall with concrete fasteners.*

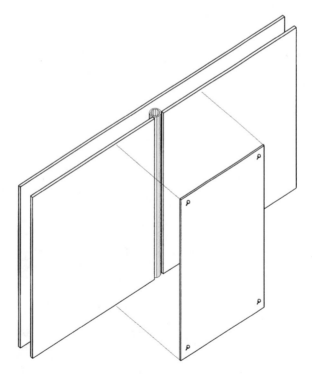

8-4 *Cutting the formwork and covering with plywood bracing to accommodate a large pipe.*

tional frame houses. Unfortunately, the tables used by HVAC contractors often do not include R-values as high and infiltration rates as low as ICF walls provide. Most experienced ICF builders prefer to install no more than two-thirds the heating Btu capacity and two-thirds the AC tonnage they would install in a frame home of similar design. (Note that these builders generally used double-pane insulating windows and above-code insulation in their roofs. In houses built with less energy-efficient features, the difference in heating and cooling requirements might be less because the greatest losses could be through the windows and roof, not the walls.) However, the final sizing decision must usually rest with the HVAC contractor, who guarantees the work.

Experienced builders are about evenly split on whether they include any air change provisions in the HVAC systems of their ICF houses. About half include a fresh air intake or an air-to-air heat exchanger to compensate for the low air infiltration. The others feel it is unnecessary. Measured air change rates for ICF houses without any other special provisions to reduce air infiltration run about .35 air changes per hour (ACH). This compares with typical rates for standard frame homes of about .7 ACH. Note, however, that air change rates vary widely with local conditions and various construction and design details. When making the decision on whether to include air change, consult with the HVAC contractor.

9

Finishes and fixtures

The nonstructural items attached to ICF walls after placing concrete include exterior siding and trim, interior wallboard or plaster, and interior trim and fixtures. All popular products can be used, but the connection details are sometimes different than they are on frame.

Exterior finishes

Troweled or painted finishes generally go onto ICFs with little advance preparation, since the systems provide a flat, continuous surface to work on. Sidings that can be screwed on (vinyl, aluminum, and so on) readily attach to systems with fastening surfaces, and even sidings requiring nailing (mostly lapped wood) can often attach directly to fastening surfaces made of plastic or wood. To screw or nail to systems without fastening surfaces, or to nail to ICFs with metal fastening surfaces, first attach nailers to the walls.

Below grade and foundation

Dampproof or waterproof the foam below grade with a foam-compatible product, as described in chapter 2. It is not necessary to have the foam surface precisely smooth or straight, but first rasp or sand any extremely sharp jags in the surface. Some systems have plastic ties with exposed ends that can cup during the pour. Also rasp or sand these as you smooth the rest of the surface, and scrape off any foam that has deteriorated in the sun. Deteriorated foam forms a yellowish powder that can pull away, reducing the adhesion of the finish.

Make sure the dampproofer completely covers any irregularities in the surface, including protruding tie ends and any gaps between them and the foam. Take extra care to seal around sleeves and wall penetrations, as you would on any basement. Extend the dampproofer a few inches above grade. When backfilling, avoid pushing large rocks against the wall. That might puncture the dampproofing membrane.

To cover the exposed foam of the foundation above grade there are different options. You can continue the dampproofing up over the foam. This is simple and economical. You can use an *elastomeric paint*, which is a thick paint with a great deal of stretch and more color options than dampproofing. You can also cover the exposed foundation foam with stucco (instructions are given later in this chapter). Stucco will be the most durable. If you are siding the house with stucco as well, you can simply continue the stucco all the way down to grade.

Preparing for above-grade siding

Before siding, sight down all exterior walls for bulges and depressions. Skin or shave the foam surface to eliminate any large irregularities. These are rare but worth checking for. If they are left, the final siding will appear uneven.

If siding with stucco, there are some extra steps. Sand down exposed tie ends that have cupped, and sand off any foam that has deteriorated in the sun. After these precautions, the stucco crew can shape the foam surface however it wants to get a final effect. Glue on other pieces of foam to make raised trim such as popouts and window casings, as desired. If you are using one of the grid panel systems, the manufacturers sell 2-inch sheets of their foam-and-cement form material to make exterior details of this sort.

Depending on the variety of stucco used, it might be necessary to cover any exposed nonfoam materials (such as lumber and plastic tie ends) with a wire mesh or a special tape for adhesion. The stucco crew should know about this.

Stucco

Stucco goes on an ICF wall just the way it goes onto frame sheathed with foam. Experienced crews are familiar with the requirements for different products.

If you are using portland cement stucco, put a continuous wire mesh over the foam surface first. If your system has fastening surfaces, screw the mesh to them with a rust-resistant screw. If not, you should have inserted U-wires in the wall before the pour, and their ends will be protruding. Twist them around the mesh to attach it. The stucco goes on at this point.

Most polymer-based stucco systems require no wire mesh. The stucco goes onto the plain foam, with a fiberglass mesh added as one of the layers. Note that if you are using a grid panel system, no wire or fiberglass mesh is necessary. The relatively great roughness and strength of the surface are sufficient to grip and support the stucco alone.

Lapped wood siding

If your system has plastic fastening surfaces that run vertically up the wall, you can install clapboard the way it is installed on frame. Nail the boards to the fastening surfaces with a ringed or hot-dipped galvanized nail. Nail the trim around openings directly to the bucks.

Occasionally, vertical pieces of trim will fall between fastening surfaces, leaving nothing to nail to. In this case, a glue directly to the foam is easiest. Once glued trim is in place, you can secure it with nails angled into adjacent clapboard or other pieces of trim.

If your system has metal fastening surfaces, plastic fastening surfaces that do not run vertically, or no fastening surfaces, you will need to install furring strips first. Put up a 1-×-2 strip every 12 or 16 inches, as shown in Fig. 9-1. To attach strips to metal fastening surfaces, use a wallboard screw. For plastic fastening surfaces, a ringed or galvanized nail is sufficient. If you have no fastening surfaces and you have installed duplex nails for furring strips, you can just pound the strips over them. Otherwise you will need to shoot a long concrete nail that goes through the foam and into the concrete, or glue the strips to the foam. For fastening trim and the ends of the clapboards at corners and around openings, install 6-inch-wide strips of ¾-inch plywood instead of furring strips. To do this job around openings, you can leave the outside flanges on your bucks, instead of adding more plywood. However, if the flanges are less than ¾ of an inch thick, you will need to build them up to that thickness.

If you want to side with wooden shingles or shakes, you will need horizontal furring strips no matter what system you are using. Space them according to the height of the shingles/shakes (usually at 6 inches oc), so that one strip falls behind each nailing line (as in Fig. 9-2). Put wider plywood strips around open-

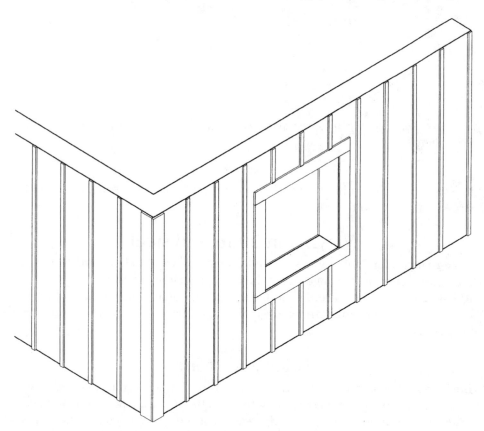

9-1 *Furring strips and plywood nailers to serve as fastening surfaces for attaching clapboard.*

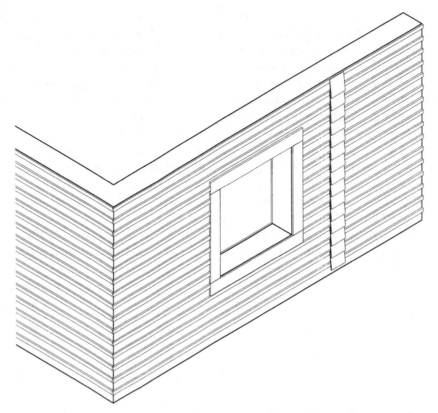

9-2 *Furring strips to serve as fastening surfaces for attaching wooden shingles.*

ings, as before, for trim. At corners you can put plywood or just continue the strips all the way to the edge.

Vinyl, aluminum, and steel

Fasten vinyl, aluminum, and steel sidings as you would clapboard (described previously). The only change is that they can be attached with screws. So if your ICF system has metal fastening surfaces, you can attach the siding to them directly, just like attaching clapboard to plastic fastening surfaces; no furring strips are necessary.

Brick and stone

Brick and stone can be installed as they are on frame—on a brick ledge with metal straps ("brick ties") connecting them to the ICF wall. Several manufacturers have special units to form a ledge easily, or they have details in their manuals that show how to make a ledge out of the standard units.

The brick ties should be installed in the foam formwork before the pour, just like U-wires for stucco mesh. This will anchor them into the concrete. If you have not done this, you can screw them into the fastening surfaces (if your system has them) or to the concrete with a concrete fastener.

Interior finish

The grid panel systems can be plastered directly. For all other systems, and for the grid panel systems if you prefer, attach some form of wallboard first.

Check your local building department before deciding what to finish with and where to put it. Many codes require a 15-minute fire barrier over plastic foam, which means using at least ½-inch gypsum wallboard. (The grid panel systems might not be subject to this rule since their foam-and-cement mixture is nearly unaffected by heat and flame.) Some local jurisdictions also require the fire barrier over ICF walls that are in some or all of the uninhabited areas of the house: the garage, basement, attic, and so on.

Preparing the surface

If you want to maximize the effectiveness of your insulation, fill any holes or gaps in the internal foam surface with adhesive foam, including plumbing and electrical chases and gaps around electrical boxes. Then sight down the interior walls and smooth out any major protrusions: bits of formwork sticking out, bulges in the wall, and so on. If you will be attaching wallboard with glue (see the following), also sand or rasp off any foam on the surface that has deteriorated in the sun.

If you have a grid panel system and you plan to plaster it directly, you are now ready to start. If you have a system with fastening surfaces, you are ready to install wallboard. If you plan to board a grid panel wall, or you are using some other ICF that has no fastening surfaces, cut a continuous chase in the foam along the top of the wall just large enough to house 1-x-2 strapping, as in Fig. 9-3. Glue the strapping into place there or fasten it to the concrete. This will allow attachment of the wallboard.

Installation

Wallboard attaches to a system with fastening surfaces just as it does to frame. Use wallboard screws and find the fastening surfaces to attach to. On nearly all systems, these surfaces are exposed or marked with a seam or ridges on the foam so they are easy to hit. In the rare systems where they are not exposed or marked, the board crew might have to hunt. Be sure to brief the crew in advance on where to find the fastening surfaces to make this easier. Also, on systems with plastic ties that protrude beyond the surface of the foam, run a continuous bead of glue around the back edge of the wallboard before attaching it, as shown in Fig. 9-4. Without this glue, there will be a slight gap between the board and the foam at the edges, so the seams could be prone to surface cracks later.

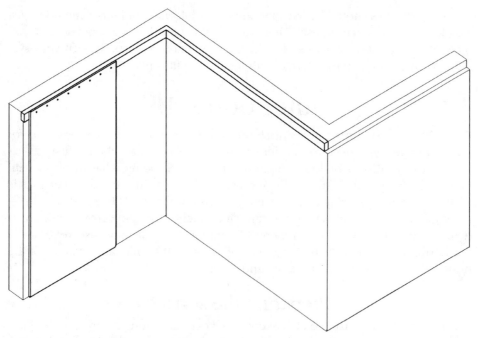

9-3 *Wooden 1-×-2 mounted along the top of the wall to serve as a fastening surface for wallboard.*

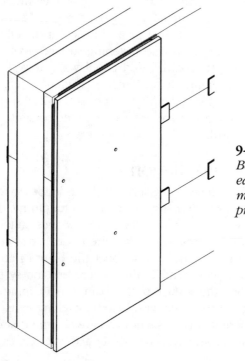

9-4
Bead of glue behind the edges of wallboard mounted on a wall with protruding tie ends.

On systems without fastening surfaces, you will glue the wallboard. Use a glue that is "foam-compatible." Others might eat away the foam rather than stick to it. Glue continuously around the perimeter of the back of each board, and spot glue in the interior. Set the boards vertically and screw them to the fastening surface you embedded into the top of the wall, as in Fig. 9-3. This will hold them securely against the wall while the glue dries.

If you are plastering directly onto a grid panel system, few special steps are necessary. Follow the manufacturer's directions.

Interior trim and fixtures

Nail the trim around openings to the bucks, just as you would nail it to the rough opening lumber of a frame wall.

Other items attach to the inside of ICF walls differently. The preferred attachment method depends on how strong a connection you need. Of course, trim or fixtures can usually be attached with a method that makes a stronger connection than they require. However, the stronger connections generally take more effort.

Lightweight connections

Soap dishes, towel racks, and other lightweight fixtures can be screwed to fastening surfaces if your system has them and they line up correctly. Otherwise, almost any wallboard fastener will work: plastic screw anchors, expansion bolts, and so on. The one variety that might not work is the fastener that depends on a low-force expansion behind the wallboard. One example is the toggle bolt, in which the "nut" is designed to expand inside the wall cavity under spring pressure. In an ICF wall, the foam will impede expansion of this nut, possibly preventing it from opening wide enough to get a firm grip.

Medium connections

Such trim as baseboard, chair rail, and cove molding can simply be attached to fastening surfaces. Remember to use a galvanized or ringed finish nail or a finish screw into plastic fastening surfaces, and a finish screw only into metal fastening surfaces.

If you have no fastening surfaces or they do not line up correctly, glue trim directly to the interior finish. You must hold the trim tight in place while the glue dries. With baseboard, press it to the wall with wooden blocks nailed to the floor. You might need to double the blocks or turn them on edge to get pressure up the entire height of the board. For moldings higher on the wall, tack in an occasional finish nail to hold the trim at the correct height, and lean scrap lumber against the trim board to keep pressure on it.

Although it is not necessary, for any trim you can also preinstall a nailer (described in the following). Some find this easier because there is no need to hold the trim in place while glue is drying.

Heavy connections

If you plan ahead of the wallboarding, you can preinstall pieces of wood on the wall and nail or screw heavier fixtures (cabinets, plumbing fixtures, built-in furniture) to them. The easiest way to do this is to take some plywood that is as thick as the interior finish (wallboard or plaster) will be. Cut a piece large enough to cover all the points you want to fasten to, but small enough to be completely hidden by the fixture when it is mounted. Put it in position on the wall before boarding and screw it to the fastening surfaces (if any) or attach to the concrete with concrete fasteners (if you have no fastening surfaces). The finish crew will butt the board or plaster up to the plywood, as in Fig. 9-5. Attach the fixture after boarding to the wood.

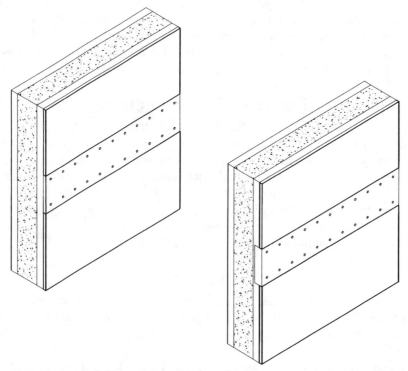

9-5 *Plywood (left) and 2-x lumber (right) mounted onto walls to serve as nailers for heavy fixtures.*

If the fixture will be particularly heavy (such as a wall-mounted sink), also glue the plywood before attaching it. Or better still, cut back the foam to the concrete. This will allow you to insert heavier lumber and fasten it to the concrete with concrete fasteners, as in Fig. 9-4. Remember to use lumber just thick enough to match the surface of the interior finish.

When installing a nailer to a post-and-beam system without fastening surfaces, you have to take into account that there is concrete only at some spots in the wall. Anchoring to the available concrete will be fine if a beam falls behind your nailer near the top, or posts fall near the left and right ends. Otherwise, the ends of the nailer will go unsupported. If such a lack of support will leave the nailer too weak for its job, make a larger nailer that extends to the next beam or posts. Make the nailer out of slightly thinner material than you would otherwise so that it does not extend out quite as far as the interior wallboard or plaster will. This will allow the interior finish crew to cover over the exposed wood with a suitable compound.

Some final details

Exterior doors ordered with extended jambs will sometimes have an extended threshhold made by connecting two ordinary-width threshholds. If your doors' threshholds have been constructed this way, put sealant in the seam. Otherwise, water can leak through and enter the house.

Along the top of the ICF walls, most of the concrete will be covered by top plates, and the rest will be exposed to the air. It is a good idea when insulating the attic to extend the insulation over the top plate and exposed concrete. Otherwise, heat can flow out of the walls through the top more readily.

Appendix

Directory of product and information sources

General information sources

Insulating Concrete Form Association
960 Harlem Ave.
#1128
Glenview, IL 60025
Phone: (708) 657-9730
Fax: (708) 657-9728

National Ready Mixed Concrete Association
Promotion Department
900 Spring St.
Silver Spring, MD 20190
Phone: (301) 587-1400
Fax: (301) 585-4219

Portland Cement Association
5420 Old Orchard Rd.
Skokie, IL 60077-1083
Phone: (708) 966-6200
Fax: (708) 966-8389

Hot knives

Demand Products Inc.
4620 South Atlanta Rd.
Smyrna, GA 30080
Phone: (404) 792-1006
Fax: (404) 792-7225

ICF systems available in U.S.
AAB

AAB Building Systems Inc.
840 Division St.
Cobourg, ON K9A 4J9
Phone: (905) 373-0004
Fax: (905) 373-0002

Amhome

Amhome U.S.A Inc.
P.O. Box 1492
Land O' Lakes, FL 34639
Phone: (813) 996-4660
Fax: (813) 996-5452

Diamond Snap-Form

AFM Corporation
P.O. Box 246
Excelsior, MN 55331
Phone: (612) 474-0809
Fax: (612) 474-2074

ENER-GRID

ENER-GRID Building Systems Inc.
6847 S. Rainbow Rd.
Buckeye, AZ 85326
Phone: (602) 386-2232
Fax: (602) 386-3298

ENERGY LOCK

ENERGY LOCK Inc.
521 West 3560 South
Salt Lake City, UT 84115
Phone: (801) 288-1199
Fax: (801) 288-1196

Featherlite

Featherlite Inc.
18 Turtle Creek Drive
Tequesta, FL 33464
Phone: (407) 575-1193
Fax: (407) 575-1959

Fold-Form

See Lite-Form

GREENBLOCK

GREENBLOCK WorldWide Corp.
P.O. Box 749
Woodland Park, CO 80866
Phone: (719) 687-0645
Fax: (719) 687-7820

I.C.E Block

North American I.C.E Block Association
4315 South Industrial Road
Suite 220
Las Vegas, NV 89103
Phone: (800) ICE-BLKS

KEEVA

KEEVA International Inc.
1854 North Acacia St.
Mesa, AZ 85213
Phone: (602) 827-9894
Fax: (602) 827-9697

Lite-Form

Lite-Form Inc.
P.O. Box 774
1210 Steuben St.
Sioux City, IA 51102
Phone: (712) 252-3704
Fax: (712) 252-3259

Polycrete

Polycrete Industries Inc.
435 rue Trans-Canada
Longueuil, Quebec J4G 2P9
Phone: (514) 646-3825
Fax: (514) 646-4880

Polysteel

American Polysteel Forms
5150-F Edith Ave.
Albuquerque, NM 87107
Phone: (800) 977-3676
Fax: (505) 345-8153

QUAD-LOCK

QUAD-LOCK Building Systems
3873 Airport Way
Suite 525
Bellingham, WA 98226
Phone: (360) 671-3911
Fax: (360) 671-7639

RASTRA

RASTRA Environmental Building Technologies
100 S. Sunrise Way
Suite 289
Palm Springs, CA 92262
Phone: (619) 778-6593
Fax: (619) 778-4917

Reddi-Form

Reddi-Form Inc.
593 Ramapo Valley Rd.
Oakland, NJ 07436
Phone: (201) 405-2030
Fax: (201) 405-1987

REWARD

3-10 INSULATED FORMS L.P.
P.O. Box 46790
Omaha, Nebraska 68128
Phone: (402) 592-7077

R-FORMS

R-FORMS/Owens-Corning
3 Century Drive
Parsippany, NJ 07054
Phone: (201) 267-1605
Fax: (210) 267-1495

SmartBlock

American Conform Industries Inc.
1820 S. Santa Fe St.
Santa Ana, CA 92705
Phone: (800) CONFORM
Fax: (714) 662-0405

Styro-Form

See Lite-Form

THERM-O-WALL

Thatcher Hill Construction
P.O. Box 593
Newport, NH 03773
Phone: (603) 863-6373

ICF systems available in Canada only
CONSULWAL

CONSULWAL
2668 Mount Albert Rd. East, RR2
Queensville, ON LOG IRO
Phone: (905) 853-2027
Fax: (905) 478-2618

I-Form

International Form Building Systems
11711 #5 Road Unit #8
Richmond, BC V7A 4E8
Phone: (604) 448-8601
Fax: (604) 448-8647

KEPSYSTEM

KEPSYSTEM Inc.
140, rue St.-Eustache
Room 301
St.-Eustache, Quebec J7R 2K9
Phone: (514) 472-3560
Fax: (514) 472-3685

Thermal cutters

Demand Products Inc.
4620 South Atlanta Rd.
Smyrna, GA 30080
Phone: (404) 792-1006
Fax: (404) 792-7225

Tri-Manufacturing
A Division of Berndorf AG
1967 Kingsview Drive
Lebanon, OH 45036
Phone: (513) 459-1300
Fax: (513) 459-9636

Windlock Corporation
680 Ben Franklin Hwy.
Birdsboro, PA 19508
Phone: (800) USA-LOCK
   or in PA: (610) 385-7436
Fax: (610) 385-4346

Pathway Systems, Inc.
P.O. Box 529
112 South Broadway
Manhattan, MT 59741
Phone: (800) 646-3243
Fax: (406) 284-6895

Thermal skinners

Demand Products Inc.
4620 South Atlanta Rd.
Smyrna, GA 30080
Phone: (404) 792-1006
Fax: (404) 792-7225

Index

Illustration page numbers are indicated in **boldface.**

About the authors

Dr. Pieter VanderWerf is an Assistant Professor at the Boston University School of Management, where he teaches courses in business and conducts research into innovations in the construction industry. He has worked as a residential carpenter and remodeler, and has worked for three years at the NAHB Research Center, where he performed business evaluations of new building products and projects. He has also worked at various times for the Information Systems divisions of Haskins and Sells and Millipore Corporation, and he has been a consultant to the Portland Cement Association, the National Concrete Masonry Association, and several private corporations.

W. Keith Munsell has 22 years of experience in general contracting and real estate sales, management, and development. He is president and founder of Resource Concepts, Inc., a construction and development firm specializing in new single-family homes and townhouses and renovations and conversions of existing properties to residential use. He is a former officer in the Army Corps of Engineers, a realtor, a licensed broker, a certified property manager, a licensed construction supervisor, and a home improvement contractor. He is also an adjunct faculty member at Boston University and Boston College, where he teaches courses in real estate finance, management, and development. Mr. Munsell was awarded Boston University's Beckwith Prize for teaching excellence.

Other Titles of Related Interest

Fireproof Homebuilding
Leo Du Lac
The only complete source available on fireproof homebuilding systems and methods.
0-07-063155-7 $40.00 Hardcover

The Portland Cement Association's Guide to Concrete Homebuilding Systems
Pieter VanderWerf and W. Keith Munsell
The first comprehensive sourcebook available on concrete-based homebuilding systems, this guide was written by two members of the Portland Cement Association—one of the major contributors to this year's New American Home, the "idea" house built specifically for the 1994 National Association of Home Builders Show. Featuring a color section of photographs of the New American Home.
0-07-067020-X $42.95 Hardcover

How to Order

 Call 1-800-822-8158
24 hours a day,
7 days a week
in U.S. and Canada

 Mail this coupon to:
McGraw-Hill, Inc.
P.O. Box 182067
Columbus, OH 43218-2607

 Fax your order to:
614-759-3644

 EMAIL
70007.1531@COMPUSERVE.COM
COMPUSERVE: GO MH

Shipping and Handling Charges

Order Amount	Within U.S.	Outside U.S.
Less than $15	$3.50	$5.50
$15.00 - $24.99	$4.00	$6.00
$25.00 - $49.99	$5.00	$7.00
$50.00 - $74.49	$6.00	$8.00
$75.00 - and up	$7.00	$9.00

EASY ORDER FORM—
SATISFACTION GUARANTEED

Ship to:

Name _____

Address _____

City/State/Zip _____

Daytime Telephone No. _____

Thank you for your order!

ITEM NO.	QUANTITY	AMT.

Method of Payment:

☐ Check or money order enclosed (payable to McGraw-Hill)

☐ DISCOVER   ☐ AMERICAN EXPRESS Cards

☐ VISA   ☐ MasterCard

Shipping & Handling charge from chart below	
Subtotal	
Please add applicable state & local sales tax	
TOTAL	

Account No. ☐☐☐☐☐☐☐☐☐☐☐☐☐☐☐☐

Signature _____ Exp. Date _____
Order invalid without signature

**In a hurry? Call 1-800-822-8158 anytime,
day or night, or visit your local bookstore.**

Key = BC95ZZA